The Reliability of
Non-destructive Inspection

The Reliability of Non-destructive Inspection

Assessing the assessment of structures under stress

M G Silk

Materials Physics and Metallurgy Division,
Harwell Laboratory

A M Stoneham   **J A G Temple**

Theoretical Physics Division,
Harwell Laboratory

Adam Hilger, Bristol

British Library Cataloguing in Publication Data

Silk, M.G.
   The reliability of non-destructive inspection:
   assessing the assessment of structures under
   stress.
   1. Non-destructive testing
   I. Title   II. Stoneham, A.M.   III. Temple,
   J.A.G.
   620.1′127   TA417.2

   ISBN 0-85274-533-8

Consultant Editor: **Mr A E Bailey**

Published under the Adam Hilger imprint by IOP Publishing Ltd
Techno House, Redcliffe Way, Bristol BS1 6NX, England

Typeset by Mathematical Composition Setters Ltd, Salisbury
Printed in Great Britain by J W Arrowsmith Ltd, Bristol

Contents

Preface

Large engineered structures are an important part of our lives. When we encounter bridges, aircraft and large buildings, or when we use electricity or gas for which pressure vessels or pipelines have been a key part of the equipment, we are relying on the successful operation of components under applied stress. How well do they survive? What margins of safety are there? The first stage, historically, was to try to build the structure and see if it lasted: trial and error at its most simple. The second stage brought in inspection, both before components were fitted and after the structure was complete. The third stage asked how good the inspection was. It is this stage which we discuss in this book.

The assessment of inspection is itself a composite of distinct contributions: quality assessment, non-destructive inspection, probabilistic fracture mechanics, combinatorial analysis, plus a range of topics from human psychology and political acceptability to the statistics of rare events. We have kept our book to modest size, resisting the temptation to cover all these topics, even at the depth of present studies (and in some areas the ideas remain at a formative stage). What we have done is to concentrate on the reliability of inspection, and here principally the detection of cracks or of flaws in components and in welds. It is just such cracks and flaws which determine the failure of engineered structures under stress. Even in this area we specialise by reference to the best documented field (namely pressure vessel studies) and the most carefully assessed approaches (principally ultrasonic). We realise, of course, that other fields (especially military ones) may have comparable data, but these are not always so accessible, so that our choice offers more data from the open literature and published assessments of reliability.

We would maintain that many of the ideas and approaches we give carry over fairly directly to other cases. Clearly, going from ultrasonic inspection to, say, radiography or magnetic methods introduces different levels of performance, but the framework of discussion is much the same. Likewise, in going from pressure vessels to networks of pipelines (as in chemical plants, perhaps, where one can by-pass units which cause concern), one needs different degrees of defect detection prior to operation. Yet the types of question one has to ask, the balances between cost of inspection and failure, the role of the operator and human factors, all have parallels.

Having said this, we note there are other fields where quite different criteria are adopted. Two extreme cases are those of semiconductor device manufacture and of medical uses of ultrasound. These are not the subject of our book, yet their differences indicate the differences of principle which can occur. In semiconductor device manufacture one can accept the rejection of large numbers of devices from a batch, and one may use methods which are destructive in a real sense, i.e. the sort of test one could not afford on a pressure vessel. Even in pressure vessel technology, of course, considerable repairs may be permissible. In medical inspection the protection of life is the aim: rejection of lives is unthinkable, and any major biopsy has to balance what it might achieve against the real dangers involved. For this reason, we have occasionally drawn attention to different approaches. The strategies, objectives and level of sophistication are quite different for medical or semiconductor device inspection. These approaches are not part of our main theme, and readers must seek fuller information elsewhere, yet they belong to the range of ideas which should be familiar to those involved in the reliability of inspection.

A further aspect of these reliability studies concerns their use. If we know with high credibility the reliability of an inspection, what do we do with this information? Here the areas of probabilistic fracture mechanics, or risk analysis, and the less tangible legal and social aspects are involved. We could not do justice to these in a single volume. Instead, we have given brief descriptions of the key notions in Chapter 2, where our aim is to give a 'flavour' of what is at issue, rather than a review. Whereas Chapter 1 discusses the background to judging approaches to assessing structural integrity, Chapter 2 discusses the context.

The main core of our book is in Chapters 3–6. Chapter 3 discusses which defects might be the topic of inspection. We concentrate on defects in steels and similar materials, though we do mention other systems which show special features. Chapter 4 concerns methods of inspection, though with an emphasis which is biased towards capability, rather than on how to do it or on popularity. In particular, one must distinguish between the ideal capability and its day-to-day reliability. These two chapters, apart from their altered emphasis, might have been the subject of a standard nondestructive testing text. Chapters 5 and 6 move away from the normal framework to discuss what might go wrong in inspection and how one can use test trials to measure, evaluate, model and improve inspection.

Chapter 5, in particular, covers the human factors and possible malfunctions of equipment. These present particular problems because they can lead to day-to-day variations in performance and because they are not easy to quantify. Yet we believe useful descriptions are possible, and that these can be assessed by idealised inspection trials like those outlined in Chapter 6.

Such studies can compare teams as well as schemes, and make a proper distinction between inherent capability and practical reliability. The recent trials concerned with the ultrasonic detection and sizing of defects in thick section steel clearly give useful information about this field of inspection. However, the trials do more than this, for they indicate the general value of this type of approach. They suggest that, despite the difficulties, such trials will become widespread as a tool in improving inspection.

We anticipate non-destructive inspection will undergo significant changes in the next few years. Social awareness and the high costs of failures are the driving forces. The means must come from systematic studies, from the increased capability of using modelling as a link between inspection and design, but above all from an informed systematic approach to reliable inspection. We hope our book can contribute to these goals.

M G Silk
A M Stoneham
J A G Temple

Acknowledgments

In any undertaking as large as writing a book, there are many who have helped the authors in at least as many ways. Our families have been quite remarkably understanding and sympathetic over what must have seemed a rather long period of preoccupation. Our colleagues have offered advice, criticism and insights. Among those from whom we have had valuable encouragement are Dr Alan Lidiard, Dr John Hudson, both personally and for his role in the PWR project from which some of the detailed studies emerged, and especially Mr Roy Sharpe, whose helpful and constructive comments on the manuscript were most useful. We have, of course, been aided by colleagues from other laboratories, notably Risley, and we should like to mention our gratitude to Mr R A Murgatroyd in relation to reliability of inspection and its human aspects, to Mr K J Cowburn in relation to ultrasonic non-destructive inspection, and to Dr W E Gardner (previously at Harwell, now at Risley) for his belief that a good theorist becomes a still better one when provoked. Finally, and by no means least, we are grateful to Miss Nicola Stoneham for compiling the index and to Mrs Jean Chanter for typing the manuscript.

All the figures are original, though seven (figures 3.6, 5.4, 5.5, 6.1, 6.3, 6.5 and 6.6) are based on figures from Harwell Reports for which the UKAEA holds copyright. We are grateful to the UKAEA for permission to publish.

1

The inspection of inspection

It is impossible to make anything foolproof, for fools are very ingenious

1.1 Introduction

At the most basic level of engineering one builds a structure, and it either does its job or (as even the advanced builders of the ancient pyramids and of the mediaeval cathedrals found to their cost) it fails. At the second level of engineering one inspects the structure and makes tests to avoid embarrassing failures. These may test the design, the component materials, or even the structure itself whilst it is being built or before use. There is a third level, in addition, in which one asks how good the tests are. Will they still allow disastrous failures of the final products? Will they be so taxing that the cost of testing exceeds the cost of failure? This book is concerned with the third level; reliability in inspection, the aims of inspection in relation to both component performance and acceptability, together with its demonstration, its monitoring and its evaluation.

The terms 'non-destructive testing' (NDT) and 'non-destructive evaluation' (NDE) were adopted during the Second World War to describe the technology of defect detection in engineering materials and components. Testing was said to be non-destructive provided that any satisfactory specimen thus examined remained fit for service after the test. Changes could occur, but not beyond a certain level. X-rays, for example, can cause changes in biological tissue, as indeed can ultrasound, yet both these techniques are regarded as 'non-destructive'. On the other hand, significant physical or chemical changes take place in the destructive testing of a component. Such destructive tests can only be used on a statistical sampling basis, and they would be wholly inappropriate for expensive, large single items such as a bridge, the pressure vessel of a nuclear reactor or even a pyramid. We continue to use 'non-destructive testing' in this widely accepted sense, and emphasise that it is inspection of stressed components which is our principal topic. This will lead us to concentrate on cracks, since cracks are usually the most serious defects for stressed components. It will also lead us to concentrate on two particular inspection techniques: radiography and ultrasonics, even though other defects and other inspection techniques are important as well and will be included in the discussions.

We concentrate on large-scale engineered structures for several reasons. They are expensive single items, and their loss could cause severe economic embarrassment or even loss of life. Such structures are usually inspected in one way or another already, and they are often the subject of legislation covering their repair and maintenance. Furthermore the cost of inspection is not usually the determining factor in the price of the structure so that (in principle at least) inspection might be carried out thoroughly. In this way we are essentially excluding a discussion of quality assurance on a production line since this usually involves the inspection or testing of a large number of less expensive items. Whilst the economic benefits of testing may well be the same or greater for a production line, the legal and social aspects of component failure are less serious in quality assurance.

In principle at least, non-destructive testing can be used to test all parts of every component. Proof testing is a halfway house between non-destructive and destructive testing. In proof testing, components are taken beyond the intended operating conditions, for example to a higher level of stress or temperature than they are expected to encounter in practice. If they survive, the components are deemed acceptable and the test is said to be non-destructive, even though the test itself could have caused some defects to grow to near-critical sizes and might contribute to any subsequent failure.

1.2   Historical attitudes to non-destructive testing

An early and familiar example of non-destructive testing is the famous experiment of Archimedes (who lived from about 287 to 212 BC) to determine the relative proportions of gold and silver in Hiero's crown. Twenty one centuries later, the tapping of railway wheels to detect cracks provides another example of the early and continuing use of non-destructive testing in an increasingly technological world. The early wheeltappers on the railway were seeking defects in the wheels and axles of the rolling stock. This was a serious business, for deterioration and breaking by fatigue of wrought iron railway wheels was responsible for thousands of accidents (Bailey 1935). Another early example, the testing of church bells, shows important general aspects of non-destructive testing, namely the nature of defects and their consequences. Savot (1627) (quoted in Singer *et al* (1957)) notes that 'bellfounders judge the quantity of tin which they should put in [the bell metal] by breaking a piece of the material before they cast it ..., because if they find the grain too large they put in more tin; and if it is too fine they augment the copper'. This relationship between destructive testing and material quality indicates the state of the art in the seventeenth century.

The roots of non-destructive testing were evident, even then. In 1664 the microscopist Henry Power (cited in Singer *et al* (1957)) recorded the first observations of metal under the microscope: 'look at the polished piece of [gold, silver, steel, copper, tin or lead] and you shall see them all full of fissures, cavities and asperities and irregularities; but least of all lead which is the closest and most compact solid body probably in all the world'. It is just these fissures and irregularities which can grow under conditions of impressed stress to yield cracks of catastrophic size.

The liberty ships of the Second World War demonstrated that welding under mass production methods was generally successful, but that catastrophic failure could occur because of embrittlement if the welds were not carried out correctly. Early welds had tended to be brittle, and welded structures of the 1930s were often suspect. This hindered the introduction of welded ships, making it a slow process, despite their advantages over riveted vessels. Welded ships were lighter and their hulls were smoother, so they could be powered more economically. Riveted construction, however, arrested cracks automatically, so that any cracks were limited in extent. The many casualties among the liberty ships were due to a combination of high stress concentrations and steel which was brittle because it was used below its ductile–brittle transition temperature. Non-destructive testing is not confined to vessels, of course. In the field of civil engineering, changes were encouraged by the effects of two world wars, the neglect of aging structures, and by the significant effect of increasing labour costs outstripping the rise in prices of steel and cement. These factors made it worthwhile to have a more precise knowledge of bridge capacity and to adopt the more economical fabrication of steelwork through electric arc welding instead of riveting such joints. Inspection techniques gained prominence because of the likelihood of defects in such welds.

In the 1930s and 1940s the importance of non-destructive testing came to be recognised. Since then, non-destructive testing, together with a greater understanding of the behaviour and failure modes of materials, has become accepted as an essential ingredient for producing reliable structures.

1.3   Recent trends in non-destructive testing

Advances in the modern technological world depend on a deeper understanding of the science and behaviour of materials, and hence on advanced manufacturing skills. These, in turn, mean that any worker sees only a small part of the manufacturing process and must rely on others both for other steps in the process and for its overall design. The old reliance on 'good workmanship' has been augmented by other techniques. One, of course, is non-destructive testing, which helps to ensure that materials enter service fit for the purpose for which they are intended, and can check whether subsequent use or abuse has allowed the initiation or growth of defects. Yet non-destructive tests would be a mere drain on resources if the tests themselves were unreliable. Too little sensitivity would allow failure to occur, and too much sensitivity would waste perfectly good specimens. Recent developments recognise how important it is to understand what makes a technique reliable, and how techniques can be improved systematically. Conversely, one should know what makes techniques unreliable, so that their limitations are clear. This book is planned as a guide to the many different factors which determine the reliability of non-destructive testing and aims to provide an assessment of the current state of the art and a framework of the main ideas and approaches for future developments.

1.4   Inspection and those who do it

Non-destructive testing can be used to find out something of interest in many different fields of endeavour. Defects which may exist in engineering structures may be detected and sized; tumours may be diagnosed or located in the human body; faults, oil-bearing geological structures, or disused mineshafts may be located underground; indeed, many other common physical measurements fall into this category. Although much of our discussion in this book could be applied to any of these fields of interest, we shall emphasise work on defects in engineering structures.

Even this is a very wide field, since the structures might be aircraft, bridges, offshore platforms, pipelines, pressure vessels, railway tracks, spacecraft or ships, amongst many other things. As mentioned earlier, most production lines making quite small components will have some form of quality control, and much of this may be based on non-destructive testing. However, we shall concentrate on large or expensive engineering structures and concern ourselves with the reliable detection of defects which could be the cause of failure. Failure will be assumed to be important because of safety or of economic factors, and the inspection is therefore particularly important. Clearly, costs incurred through poor quality control on a production line are important too, and much of what we discuss will probably apply to this situation as well. Our reason for concentrating on large, expensive items is that the level of non-destructive examination which can be afforded, in terms of both the cost of sophisticated inspection equipment and the time allowable, places less restriction on thorough inspection.

As we have remarked, the reliability of non-destructive testing depends upon a very wide range of factors and we shall discuss the nature and interdependence of these in the following chapters. Some of these factors, such as the nature of the structure under examination, the types of defect being sought and the choice of the non-destructive testing technique to be employed, are dominated by scientific and engineering considerations, and will be discussed in these terms. Other aspects, such as the reliability of inspection equipment, the ergonomics of the use of equipment on site and the performance of the human tester, include physiological and psychological factors. Understanding such factors is a separate subject, and in the following discussion we have to rely on the statistical information which exists in the literature. Moreover, the data on some of these human factors must be obtained from fields of study other than non-destructive testing.

To give one example of this, it seems clear that factors such as boredom must play a part in the day-to-day performance of an inspector. It would not be surprising to find that the reliability of defect detection may be related to the number and nature of the indications which the operator has seen—or has not seen—in the previous work period. Yet none of the round-robin tests so far envisaged for non-destructive testing could lull operators into the conviction that they themselves are not being tested, and, in any case, such tests rarely include specimens without defects. For this reason an operator is likely to remain highly motivated and interested throughout the period of the test both by the knowledge that his or her results are important and by the regular occurrence of significant indications. Any quantitative estimate of the performance of NDT operators doing their day-to-day tasks must be extrapolated from other sources of data, when the presence of significant defects is a rare occurrence. Clearly, one must distinguish between the *capability* of a technique, i.e. its inherent design limitations, and its *reliability* in day-to-day practice.

A final factor to be considered arises from any attempt to combine the effect of these sources of unreliability to produce, for a given structure, an estimate of the probability of failure. This involves the use of statistics at quite a complex level, since it is doubtful whether the individual contributing factors can be regarded as being independent. In addition, since the failure of important structures is an unusual event, the estimates of the probability of failure will be based on the 'tails' of the assumed probability distributions. Clearly one issue concerns the way one tries to parametrise an existing database to make predictions of safety levels. The quantification of the data in these tails will be based less rigorously on theory and, at the same time, can less easily be confirmed by statistical tests. The effect of the superposition of errors in the estimates of reliability must also be considered in making any overall estimate.

1.5 General plan of the book

In the next chapter we consider the effect of the non-destructive testing techniques and procedures on the reliability of the system being inspected.

In this discussion the ways in which the individual sources of reliability impinge on the overall reliability are considered. This leads on to a consideration of the approaches used in the analysis of risk where once again we find that human factors may be important, in this case in the extra factor of the perception of risk. We have included too an outline of some of the relevant statistical methods and theories.

The basic contributions to the reliability of non-destructive testing arise from the nature of the structure itself, the environment in which the structure is expected to operate, and the kinds of defects expected to occur. From this information it is possible to estimate which are the defect types which might be expected to lead to failure and the maximum size of these defects which can be tolerated. The considerations which are important in this context are discussed in Chapter 3 and, from a consideration of these, the basis of the non-destructive testing requirements begins to emerge.

Armed with an estimate of the size and nature of the defect types which could lead to failure the next step would be the choice of a suitable non-destructive testing technique. The basis of this choice should be the capabilities of the techniques themselves which involves understanding the way in which defect detection is accomplished in each case. This may seem an unexceptional statement but it is surprisingly common to find (say) radiography and ultrasonics being considered as alternatives which can be fully substituted for one another, without any consideration of their strengths and weaknesses. Of course, when more than one technique is capable of performing the inspection, the choice may include further factors such as convenience or existing equipment, but this choice should also take into account a due consideration of aspects discussed later in the book. In Chapter 4 we provide a brief review of the most commonly used non-destructive testing techniques, the basis on which they work and the types of defects which they are likely to be most successful in detecting.

In Chapter 5 we present a detailed consideration of the reasons for the failure of inspection. In addition to the adequacy of the chosen inspection technique we introduce those effects which depend on the human operator, for over 95% of current non-destructive testing is still carried out manually. For example, we may ask whether an automated scan is an improvement on manual scanning. What influence do testing procedures have on reliability? What kinds of errors may operators themselves introduce? Another contributor to unreliability which we consider is that of the equipment itself. This may affect operator performance if it continually fails itself. This is especially true since instruments have been produced in which failure is not immediately detectable—a 'null' result is expected from a material without defects. Finally we consider the effects on otherwise suitable inspection techniques of specific material or defect properties and the influence of the point-to-point variations in these properties.

A major contribution to the reliability of non-destructive testing is the calibration of the techniques. This may either take the form of the use of standard calibration specimens, such as the calibration blocks used for ultrasonics, or the use of representative samples cut from the material being inspected (or of the same type as that being inspected). Ideally, the latter would include representative defects, but this is very often not possible; in the majority of cases a series of agreed artificial defects is introduced. Going one step further than calibration is the direct trial of a technique or a piece of equipment using specially manufactured or selected test blocks. Here the nature of the defects in the specimen is unknown to the tester. These tests range from a simple trial, aimed at a single inspection requirement, to the recent wide-ranging national and international trials aimed at assessing the capability of certain ultrasonic inspection techniques. We discuss all of these approaches in Chapter 6, where we consider the usefulness and pitfalls of these attempts to quantify reliability.

Quantification can, of course, take several forms and these depend on how reliability is defined. One might choose the fraction of defects missed, or misgraded, or one could choose a hazard threshold which depends on the consequences of failure.

There is a final term which we must mention in this introduction. This is the influence of financial considerations on the reliability of non-destructive testing. Our emphasis in this book on the larger expensive structures is largely because of the difficulty of including the effects of any financial limitations in any quantitative way. Nevertheless it is important to note that it is normally the case that the potential costs of the inspection of a structure are treated in a different way to those of its design, construction and insurance. This possibly results from a feeling that non-destructive testing is unreliable or a belief that insurance and non-destructive testing are fully interchangeable alternatives. Unless the use of non-destructive testing can be shown to be (or at least believed to be) of direct benefit, there is apparently little incentive to treat it seriously unless statutory requirements demand it. That this is so is surprising. One might have thought that reputation would provide a powerful incentive, since one cannot insure against a damaged reputation. However, the short-term effect is often that cost, rather than reliability, may be the overriding factor in the choice of an inspection technique, even though a good quality inspection might be expected to save both long-term operating costs and allow insurance premiums to be reduced.

Another disappointing factor is that there is still little feedback between inspection and design. Often fairly trivial design changes can reduce the costs dramatically by their effect on the reliability of inspection. These changes might substantially reduce the overall cost of the structure to the user, particularly when a structure has statutory inspection requirements. Clearly cash-flow considerations and the common distinction between capital and running costs play a large part in determining the economics of these approaches. Even so there appear to be opportunities for improving the reliability of non-destructive testing. Reliable non-destructive testing, in our view, is not an expensive luxury, but a potential source of greater economy, efficiency and safety.

2

Systems reliability

Investment in reliability will increase until it exceeds the probable cost of error

2.1 Introduction

In this chapter, we take a step back from NDT methods and practices to look at the context in which inspection is done. We examine the importance of specific tests for the success (and especially the safety) of a system as a whole. We shall also discuss the more general legal and social aspects, though these warrant a separate book to themselves. It is now common for complex advanced engineered systems to receive keen and critical scrutiny of the way they work and the ways in which the consequences of their failure are handled. This scrutiny is often in the name of public safety, rather than commercial cost, and may be based on past experience (like aircraft crashes), forseeable experience (like dam failures) or on purely hypothetical combinations of incidents (as in many of the probabilistic risk studies carried out for nuclear plants). Clearly, the risk which is associated with these systems contains several distinct components.

We split our discussion into several parts. First we look at the general ideas about *failure*. This includes a range of ideas, including lack of availability, unfitness for the chosen purpose, whether or not replacement is possible, and the consequences of the failure. *Fitness for purpose* proves an important concept in NDT specification. Failure also involves the systematic relation of *component* failure and *system* failure. The organised analysis of the relationship is a necessary (though not sufficient) step in establishing a degree of reliability for a system, however defined, and in optimising an inspection strategy. Risk analysis and probabilistic methods are important here.

2.2 Fitness for purpose

The specification for some stressed component, or even for a whole plant,

9

can be expressed in two quite extreme ways. The first, which one could term the 'Elysian' approach, is to seek perfection in fabrication. It proposes that one should aim for the best quality achievable with existing methods, irrespective of final application. The corresponding code might read: (I) 'No defects'.

The second approach recognises the variations in significance of defects from one type to another. The relevant quality is the extent to which the defect affects operating efficiency. A code in this 'fitness for purpose' category might read: (II) 'No cracks larger than C; no regions of lack of fusion larger than F; defined acceptance levels of porosity or slag.'

In principle, other types of specification are possible. One might define a performance standard, for example that a component will survive a defined loading. The inspection agency (rather than the manufacturer of the final product) would then have to define a code such as II. The inspection agency may even wish to supplement the tests by methods such as hydro-testing which may possibly be destructive. With such performance specifications contractual responsibilities are changed, and we comment on this in §2.5. For the present we discuss only the two extreme specifications such as I and II.

Fitness-for-purpose specifications give major advantages over Elysian options (Wells 1982, Farrer 1982). First, there is an effect on cost and on delivery schedule. Not only will apparent perfection take longer, and be more expensive to achieve, but it will be more unpredictable. One cannot use past performance as a guide, so that wider margins of cost and delays will be needed. Secondly, times taken for repair with Elysian specifications are not easily predicted. Repairs of what might be irrelevant defects may take several attempts and achieve frustration of the welder (and hence probably undesirable human factors) without useful benefit to the product. Indeed, attempts to remove harmless levels of porosity could make matters worse by introducing lack of fusion in the repair welds. Thirdly, there may exist no known standard technique to marry to the Elysian standards. The procedures may only emerge (if they do at all) by much trial and error. With fitness for purpose, there will usually be readily identified means of producing and verifying the defined end, and repair needs can be modest when proper training methods are adopted. Fourthly, fitness for purpose can have clearly defined NDT codes, without the element of subjective interpretation in the Elysian case. Finally, Elysian methods may even be harmful, for they give an illusion of extreme operating safety which is based mainly on reinforcing aspects already safe, without necessarily excluding other mechanisms which may possibly be serious. This leads to the systematic approaches outlined in §2.3.

Whilst the spirit of fitness-for-purpose specifications has clear merit, it needs a substantial scientific basis and database. The materials used, like the grades of steel, weld materials, heat treatments etc, influence both performance and cost. Knowledge of defect detectability and influence on risk and on the cost of failure is needed. Moreover, the technical control (covering supervision and definition) of subcontracted work has to be managed and can be hard to achieve. There can also be management problems.

2.3   Analysis of system failure

We now turn to systematic methods of looking at the possible failure of some complex system, whether it be a nuclear reactor, chemical plant or air-craft. The aim of these analyses is to reduce costs from the time that the plant is out of service, or to minimise risk to users and the public at large, or some combination of the two. Our aim here is to give a flavour of the way such analyses are used to improve the reliability of inspection and to show where data on inspection reliability might be used. For detailed prescriptions on how to carry out these analyses the reader should consult one of the standard reliability engineering textbooks such as O'Connor (1981), Green and Bourne (1972), Billington and Allan (1983) and the Open University (1976).

2.3.1   Fault and event trees

The two common methods of analysis are those of the *fault tree* and of the *event tree*. Both use probabilistic arguments, and both aim to relate im-probable system events to more probable component events. In practice, both approaches are managed by general computer codes based on standard mathematical methods such as set theory.

The *fault tree* is especially useful in discussions of routine operation and maintenance. It tries to analyse from effect to cause. One identifies an event (called the top event, like 'the plant stops working') and examines how it might have happened in terms of the components (the primary events) which could have contributed to the subsequent top event. The fault tree (for example figures 2.1, 2.2) is simply a graphical way of displaying how failure of a system can result from failures of components. Only faults are included, i.e. a non-failure is simply omitted. Time dependence is included, because failures may not be immediate, but the way that the time dependence is included is more combinatorial that dynamical, i.e. the probabilities of various delays are recognised but the blow-by-blow treat-ment of forces and responses is not carried out within the programme. The limits of fault tree methods are partly intrinsic and partly practical. As described, a component either works or fails, and it is always in one of these two extreme states, whereas the undoubted intermediate cases are not given separate treatment. Moreover, contributing events need not be as independent as presumed. Secondary failures, outside the range of behaviour considered explicitly, might limit the success of repairs. More importantly, there can be *common-mode failures* (see §2.3.2) where several failures arise from a common cause, and this is harder to represent. On the practical side, completeness is difficult to achieve; if achieved, the result may be too complex to interpret readily; even if interpretable, it may rely heavily on inexact estimates of the reliability of specific components.

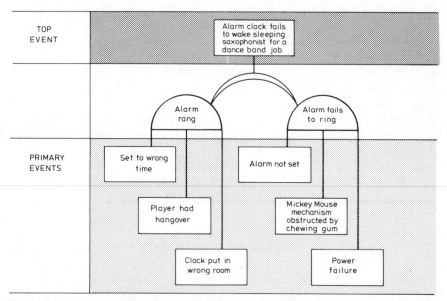

Figure 2.1 Simplified fault tree. The event tree version of the same in events is shown in figure 2.3

We may take a simple example in which a severe defect is not reported during an ultrasonic inspection (see figure 2.2). This is the top event. The fault tree identifies two possible contributing causes: either the data-recording equipment has failed, or no interrogating signal was injected into the component. In the case that the signals were not injected into the component, two possible causes are identified: either the transducer did not fire or there was a loss of coupling between the transducer and the component. For this example these two contributing low-level events have been assigned probabilities P_f and P_c respectively. The other two low-level events, contributing to the other branch of the fault tree, combine to cause failure of the recording equipment. These events are that both receiver A and receiver B fail together and have been assigned probabilities P_a and P_b respectively.

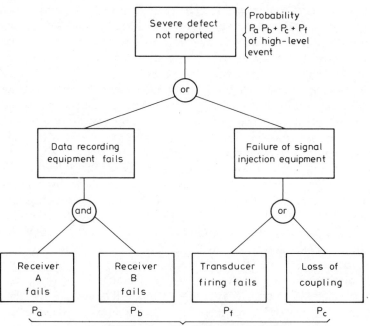

Figure 2.2 A simplified fault tree for a non-destructive test showing how probabilities of low-level events can be combined to estimate the probability of a high-level, or serious, event.

Since both these low-level events must occur together we write $P_a P_b$ for their joint occurrence and add the other two low-level events P_c and P_f. Note that these probabilities should not be combined numerically until the final step is reached, as the variables are essentially Boolean in nature and are not actual probabilities until some Boolean algebra has been carried out. The Boolean algebraic manipulations give, for example, $P_a^2 = P_a$ and $P_a + P_a = P_a$ (for more details see Kinchin (1979) and Unwin (1984)). In our simple example, provided that the individual low-level events are independent and unlikely, the overall probability of the top event is given numerically by $P_a P_b + P_c + P_f$. The assignment of numerical values of probability to the low-level events can be achieved by carrying out experiments, by analysis of operating experience or sometimes simply by guessing.

The *event tree* is particularly useful for analysing emergency situations. Here the analysis proceeds from cause to effect. One identifies an *initiating event* (or stimulus) and looks at the way the system responds. The result is really a summary of the consequences of a particular event. (There are usually no points at which the outcome is influenced by a decision.) The limitations of event trees lie partly in the difficulty of organising the time groupings of very rapid events and partly in the possible scientific uncertainties in behaviour. Event trees are shown in figure 2.3 and 2.4, showing the way probabilities must be assigned and propagated to obtain the overall probability of failure. For details of computer codes to carry out the analyses of large event trees see Hall (1984).

In figure 2.4, we have set out a simplified event tree where the initiating event has frequency F. Branches going to the right and upwards represent cases where errors are trapped, whereas a downward step at any stage allows errors to propagate. Each time an error is trapped with probability P it propagates untrapped with probability $1 - P$. We have shown an event tree with three levels of protection. Error trapping at each level occurs with probability P_1, P_2 and P_3. The number of times the system is fully protected is given by $FP_1P_2P_3$ whilst all the system error traps fail with frequency $F(1 - P)(1 - P_2)(1 - P_3)$. Thus if, as is likely with many error-trapping mechanisms, the probability of success at level 1 is 9999 in 10 000, at level 2 is 999 in 1000 and at level 3 is 99 out of 100 occurrences we would assign values of $P_1 = 0.9999$, $P_2 = 0.999$ and $P_3 = 0.99$. So, of F occurrences of the initiating event, $F \times 10^{-9}$ will manage to by-pass all error-trapping mechanisms and 98.89% of F events will be successfully dealt with by the system. A whole spectrum of intermediate results is also possible, as indicated in the figure.

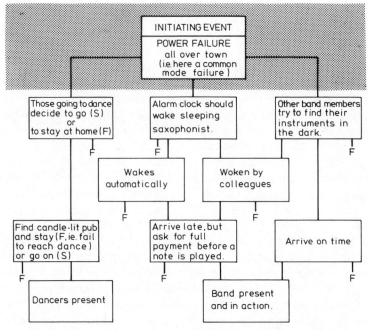

Figure 2.3 Simplified event tree corresponding to figure 2.1

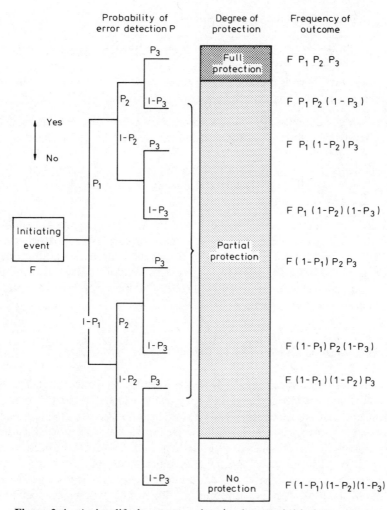

Figure 2.4 A simplified event tree showing how an initiating event can be traced to a series of possible outcomes each with an associated probability.

2.3.2 Common-mode failure

The overall probability of failure can be enhanced by failures in several separate ways from some common cause. More precisely, a common-mode failure means that the probability of two or more components failing is higher than if all the components behaved independently. Estimates— usually attempted from existing databases—are often expressed as a β-factor, describing the fraction of the total failure rate attributed to dependent failures. We return to common-mode failures in §2.4.6.

2.3.3   Regeneration and repair strategies

When a component fails, there are several different approaches to the replacement. Clearly the replacement must work—it would not be a repair otherwise—but there is a choice depending on how far through its lifetime the replacement is. Some of the standard cases are these.

Better than before: a system is said to be *coherent* if its performance is always improved when a component is improved. The component is then said to be *relevant*.
As good as new: e.g. replace a worn-out battery by a brand new one.
As bad as old: e.g. buy a working battery from a car dump.
Not quite as bad as old, etc ....

Regeneration diagrams are used to estimate the system availability achievable following these strategies. Repair strategies are not necessarily given by the *failure* analyses we have discussed. The case of time out of service is one of probabilistic economics, for it involves just the combination of finance and estimates of likelihood. The methods parallel actuarial approaches to life insurance.

2.4   Risk analysis and human factors

The previous section discussed how one could organise the often vast amount of material describing failure mechanisms. The final result was pictorial rather than quantitative, and sometimes made use of ill-defined or incompletely known relationships between individual events. Risk to users and to the public raises issues of public acceptance and corresponding moral questions. We now move on to how one can calculate probabilities, usually so that some non-specialist can judge whether what has been done is really acceptable. Probabilities are unpopular with non-sporting non-specialists, however, so we look first at ways which might minimise the unpopular aspects.

There is nothing so uncertain as human life, it is said, yet nothing is so sure as a life insurance society's profits. So why isn't it possible to use actuarial methods? The answer is lack of recorded experience. Life tables, listing longevity, accidental death and the many influences of mortality, have been assembled systematically. Recorded experience of major accidents with modern technology may be zero, for any serious accident usually leads to changed designs—and there may never have been a serious accident. As noted elsewhere in this book, one cannot do *statistics* when the averages are grossly changed by excluding a single datum.

The problems are clear. Human failings are not deterministic: there is always another wrong way. Acts of God (or strictly, acts not under human control) are only predictable with limited warning (the typhoon is nigh; the earthquake tremors still the birdsong (the reverse of acoustic emission!)) or on a probabilistic basis. Even if these two disadvantages are insufficient, there are usually uncontrollable factors outside the system one would like to treat deterministically.

If one has to use probabilities, cannot their own uncertainties be reduced by looking only at the worst option at each stage? Sadly this is often inaccurate to the extent of being ridiculously over-pessimistic. There may be no useful worst case; for example it may always correspond to system failure as it is switched on. This depressing result may owe nothing to common sense (or, indeed, any other form of sense) but all to some limitation of the database.

The limitations of the actuary's database approach lead to two strategies to combine what is known of reliability to estimate the impact of its failure. In this context the risk has a formal definition as the product of P_F, the failure probability, and S, the severity of the failure, should it happen:

$$\text{risk} = P_F \, S$$

and the two factors are examined separately.

2.4.1 Probabilistic engineering analysis: probabilistic fracture mechanics

This strategy uses deterministic models, e.g. stress analysis, which relate lifetime (or similar parameters characterising failure) to the controllable engineering parameters, e.g. materials specifications, tolerances, etc. Each model parameter has an assumed statistical variation which is included in estimates of P for the system as a whole. There are, in fact, several approaches and ways of presenting results. Nevertheless, it is clear that three separate probability distributions are involved, namely

(i)   the probability that a specific defect of specific size will give failure;
(ii)   the probability that inspection will succeed for those defects of significance;
(iii) the probability that there is a flaw in the material or component prior to inspection.

If we wish to base a component-replacement strategy on probabilistic methods, we shall need probabilistic information about the defect evolution as well. Our own concern in this book is with (ii) as discussed in §6.9, and it is useful to remark on inspection success as more methods are tried. We note that the better the data for the several probability distributions, the less is the need for the expensive, and often unhelpful, safety factors included to cover ignorance.

2.4.2 Combined analysis: engineering and statistical aspects of field experience

Here one calibrates an (incomplete) engineering model against available data from the actual service operation of the system. Clearly the engineering model needs to be formulated with the right degree of experience and judgment. Once this is done, the approach resembles regression analysis. The reliability of pre-inspection is not exploited to any great extent, but the inspection of failed components is, of course, a source of operational data.

2.4.3 Human aspects

This is not a book of applied psychology. Yet there are several places where subjective criteria are involved, and it is useful to bring them together in this section. Wherever possible the topics we shall cover are ones where a *subjective* assessment can be compared with an actual and quantitative *objective* value. Some aspects are concerned with management within one's own organisation. How do you make sure inspectors will report faults without succumbing to moral pressure from the production staff? How can one prevent botched repairs being made because a proper job would threaten schedules? Other aspects are concerned with design: 'Although Heath Robinson died some years ago, his disciples are still around and employed by some contractors' (Kletz 1983). The analysis after the event of the Three Mile Island incident showed how easily display format could affect operator response and so influence what happened. Design for efficient function is possible, and effective designs demonstrate this by their persistence. The survival of the QWERTY typewriter keyboard format shows that its design is of long-term value (see Salthouse (1984) who discusses still better options).

2.4.4 Estimates by consensus: the Delphi technique

Here we are concerned with the views of an 'expert' in the field; i.e. the informed and experienced person whose professional needs include good judgment. Green (1983) provides a number of examples. The conditions of these estimates of failure rates for various items of equipment were:

(i)   those who made the estimates were a cross section of experienced engineers in the field;

(ii)   the equipment could be seen, inspected, and diagrams and manuals were available;

(iii) all equipment had been used in the field with good sample sizes and operating experience—though, clearly, the individual engineers may have only had personal knowledge of a fraction of that. (There was, however, a severe time limit on reaching decisions.)

The results cited by Green refer to electronic equipment and mechanical equipment similar to that in nuclear protective systems, and to various airline systems. Whilst the precise figures (see tables 4.1–4.5 of Green (1983)) are not directly relevant here, they show three relevant features. First, there is wide scatter. Individual estimates could easily be out by two orders of magnitude either way. Secondly, the average estimates were very much better: most to within a factor of three, and all to within one order of magnitude. Thirdly, the distributions of estimates were skewed and roughly log normal (i.e. the distribution of the logarithm of the failure rate was roughly a normal distribution): people 'tend to estimate logarithmically'. This 'Delphic' approach seems to offer, therefore, a useful starting point for preliminary analyses, and is probably an unavoidable method in the absence of large databases.

2.4.5 Subjective risk

Risk comes into the reliability of non-destructive testing because it governs the criteria which are set. Major disasters, like the Tay Bridge (see *The Innovative Engineer* 1981) have had their impact on validation and statutory tests, and properly so. It is equally well known that perceived risk and real risk are not simply related (see, for example, Fischoff *et al* 1978, Lord Rothschild 1978, Slovic *et al* 1979, Reisland and Harries 1979). In any attempt to make life safer, priorities have to be determined between reducing public dangers and reducing public fears.

The subjective components of risk take several forms. One is voluntary versus involuntary exposure, often taken as employee risk versus public risk. Another is the size of the event which causes casualties, though the known consequences of massive dam failures (such as the tens of thousands drowned following the Morvi dam failure) do not seem to produce widespread concern. Yet again, the categories of death have subjective components: real, immediate deaths (e.g. run over by a steamroller), hastened deaths (e.g. from the effects of shock, having seen someone else run over by a steamroller) and the old favourite of the anti-nuclear lobby, hypothetical deaths (those who might have been run over by a steamroller, but weren't). Familiarity is another factor. Travel by car, it seems, has dangers underestimated in comparison with flying and overestimated in relation to being employed as a firefighter.

2.4.6 Human factors: statistics and subjectivity

Kinchin (1979) has described how to evaluate the probability that equipment failure or human error will lead to an unreliable inspection. Event tree or fault tree analysis may be used. As remarked earlier, for event tree analysis we would start with the initiating event and trace the course of all possible sequences leading to the culmination of a defect incorrectly accepted when, in fact, the inspection procedure would have required rejection of the component as unsuitable or unsafe, say. Simplified event trees relevant to this discussion are shown in figures 2.3 and 2.4. It is just as easy to work back from the final result, the unreliable inspection, and create a fault tree. This traces the outcome through failures of subsystems and down to a level of component failures for which data are available. Simplified relevant fault trees are shown in figures 2.1 and 2.2. Common-mode failures are important causes of unreliable inspections. These are mentioned in §2.3.2.

Here we note that a cause of common-mode failure can often be the human operator. To include the effects of this in an analysis of fault trees or event trees it is necessary to determine the degree of dependence between some of the responses and the situations in which the operator may be asked to perform. Coupled actions precipitating common-mode failure can be calculated (Edwards and Watson 1979) by assuming an effect which is intermediate between that produced assuming complete independence of operator actions A_1 and A_2 with probabilities $P(A_1)$ and $P(A_2)$ and that assuming none. Mathematically, we express this as:

$$P(A_1) \, P(A_2) \leqslant P(A_1 A_2) \leqslant P(A_1) + P(A_2)$$

an inequality which can be achieved, for example, by the geometrical mean:

$$P(A_1 A_2) = [P(A_1) \, P(A) \, (P(A_1) + P(A_2))]^{1/2}.$$

Alternatively, we can introduce a time dependence into the operator actions such that if an error has been committed j times in succession then the probability that it will be repeated is a non-decreasing function, and if the actions are widely separated in time they are uncorrelated. Taking γ_0 as the probability of operator errors for the first time and γ_j, where $j = 1 \ldots n$ as the conditional probability of human error repeated for the $j + 1$th time, then the probability that two facets of the inspection are in error due to human mistakes is P_r with: $P_r = \gamma_0 \gamma_1 \ldots \gamma_j$. If the actions are widely separated in time we have $\gamma_0 = \gamma_1 = \gamma_2 \ldots$, but when carried out in succession under a high level of stress,

$$\gamma_j \rightarrow 1$$

and generally, $\gamma_0 \leqslant \gamma_j \leqslant 1$ (Edwards and Watson 1979). In §5.5 we present a figure of between 10^{-4} and 10^{-3} for the rate of serious human error.

It is difficult to validate human reliability assessment techniques, as Williams (1985) has pointed out. He notes that potential users of human reliability data appear to be most sensitive to high and low predicted error probabilities. High error probabilities may imply a need to build in greater diversity or redundancy. Low probabilities cause concern because users experience difficulty believing in lower error predictions. Only limited attempts have so far been made to validate the predictive capability of published assessment techniques. Kirwan (1982), from a study of three subjective human reliability techniques, concluded that 'absolute probability judgement' (i.e. *guessing* (Williams 1985)), when compared with the known probability for an event, would be accurate to within a factor of two on about 30% of occasions, correct to within a factor of four on about 60% of occasions, and almost always accurate to within a factor of ten. The study also concluded that no expert or person trained in human factor or reliability analysis performed significantly better than any other. Although some individuals did consistently perform better than others, the precisions achieved were not markedly different to those achieved with group median responses.

2.5 Legal and economic aspects

It was with some hesitation that we decided we would also include some legal and economic aspects of the reliability of non-destructive testing. Provided the inspection is done with reasonable attention and effort, it seems at first that almost all responsibility for subsequent disasters can be passed to someone else. It is the manufacturer or the person who writes the specification of equipment who decides which defects are to be checked. There is a code of practice which gives rules for inspection. The retailer—or perhaps the manufacturer—is the one who receives initial customer complaints. Moreover, even if all goes wrong and the blame is clearly attributed to an undetected significant defect, it is rare for any legal protection for the inspector to be necessary except perhaps a sensible contract and general liability insurance.

Despite all these limitations, we felt a discussion of some of the issues would be helpful since there seems to be no summary of these issues directly relevant to inspection. Many of the points we discuss will be treated at a superficial level, and we emphasise that the reader will certainly have to seek specialised advice at times of need.

2.5.1 *What is a reasonable contract?*

Contracts contain several types of provision listed below. Some are common to all, whether selling a used car or designing a space satellite. Others are specific to the area of contract work, including inspection.

(*a*) General legal provisions. These include the legal titles of the agencies making the contract, the dates for which it is in force, and which law is to be presumed in interpreting it. For example, in international contracts, it is usual to insist that the law of one named country will apply.

(b) Financial. These provisions include the dates for submission of invoices, methods and dates of payment, and the currency of payment. There should also be procedures outlined for the action needed if overrunning or overspending seem likely.

(c) What do you promise to do by way of inspection? Here it is sensible to promise what you will *do* rather than what you will *achieve*. Thus it is realistic to say N operators will spend D days or examine S samples according to a particular code. It is unrealistic to contract to find *all* defects within a specification for reasons obvious from the rest of the book.

(d) What should be provided by the manufacturer of the component tested? First of all, you should have specimens. Any dates agreed (see (a) above) should make it clear that you cannot start work without samples, and that delays in their arrival should allow you to delay completion without penalty. Secondly, you should have a specification. This can contain three main elements, namely the defects which are sought, the methods to be adopted, and sometimes a decision about the safety or otherwise of the defects detected and any action necessary. It is at just this point that questions of responsibility become blurred, since a manufacturer may often seek the advice (whether formal or informal) of the agency carrying out the inspection.

It is sensible to make the manufacturer responsible for identifying the defects to be sought, since he will usually be held responsible for consequences of the defects. Some standard cases are covered by engineering codes of practice. In other cases, those doing the inspection may draw attention to any obvious shortcomings (preferably in writing!). However, if the situation is really a novel one, it makes sense for any work defining a specification to be done on a separate contract from the actual one defining inspection work.

The methods to be adopted are again often covered by long-established (or at least accepted) codes of practice. Some are old and arguably obsolete, but they do provide a minimum performance, if occasionally for maximum effort. If new approaches are to be tried, then demonstration and formal confirmation of acceptability are advisable.

Whether or not a defect detected is acceptable (i.e. leaves the component fit for service) can be a problem, especially if there is any uncertainty in the specification. Since manufacturers prefer to believe their products to be free of defects, there may be encouragement to ignore those borderline cases which the manufacturer would prefer to forget. The responsibility is clear: the person carrying out this inspection should report marginal defects and the decision to take action or not rests with the manufacturer who, after all, presumably can check seriousness with those who drew up the original specification. The issues can be quite subtle, since 'defect size' interpreted from a chosen inspection method may differ, even for the same defect, from that used in any design based on linear elastic fracture mechanics.

(e) What is to be done with the results? Inspection involves two distinct aspects, namely the specific results (like ultrasonic signal records) on an identifiable set of samples, and the know-how associated with methods used to take, analyse and interpret data. As a rule, the specific results will be the manufacturer's property. If you want to use them in technical articles, in advertising material, or in related contracts, you should have formal agreement. The methods you have used may be more general, possibly widely applied. If any development work is done, possibly no more than simple computer codes for automation of a simple step, it is important to have defined the extent of your freedom to use the method in other contracts. You do not want an omission on a small contract for racing-yacht keels to destroy your chances of massive contracts for supersonic jet wings. This can be especially frustrating if the methods have been around for many years. A proper clause for the intellectual property is of real importance.

2.5.2   Social and moral aspects

The reliability of non-destructive testing is only one part of the definition and achievement of a safe system, whether that system be a bridge, a pipeline, a reactor, or an integrated circuit controlling a manufacturing process. Nevertheless, we are surely right to discuss broader issues, like how safe is indeed safe enough?

The aim is, of course, to take the best of all possible actions (actions here include doing nothing, and also delaying taking decisions, i.e. inaction is an action!). But best is not uniquely defined. Schulze (1980) has classified some of the possibilities:

Scheme	Aim
Utilitarian (Benthamite)	Greatest good of the greatest number
Egalitarian (Rawls, Kant)	Well being of the worst-off
Elitist (Nietzschist)	Well being of some elite
Libertarian (Pareto, Christian)	Any change leaves no-one worse off
Cost/benefit	Balance between cost and achieved benefit.

At any time, choices among these ethical schemes may be limited by social or practical constraints. At the time of writing, for example, it is hard to imagine elitist schemes finding favour in Western Europe.

What is practical is itself a field for both informal belief and legal definition. For civil actions, two relevant legal definitions (Redgrave's Health and

Safety in Factories 1976) are those of *reasonably practicable* and of *practicable means*. These terms imply a balance to be made between the quantum of risk and the sacrifice (i.e. cost in money, time or trouble) in measures necessary to avert the risk. In the event of a complaint, the onus is on the person responsible for such measures to prove what compliance was not reasonably practicable, the computation of the balance to be made prior to the incident complained of. These two terms are obviously qualifying phrases, i.e. unlike those legal obligations which are absolute, imposing liabilities even though the person liable had made extensive efforts to avoid the calamity involved (see Barton (1982) for example).

2.5.3   Legal aspects: general points

Our discussion of contract provisions raises a number of general issues, which we bring together here.

First, there are various defined levels of responsibility. Some are defined by statute, and are specific provisions of the law of the land. Some are given by well defined, widely accepted codes of conduct or of design (e.g. British Standards or ASME). Others are covered by the contract itself, or by common law. The formal legal standards do offer protection against a priori irresponsibility. At other levels the argument is less clear, and the definitions and degrees of proof are those of law (especially case law) rather than of a scientific proof or technical definition.

Secondly, no general protection can be guaranteed if a non-destructive inspection is neither non-destructive nor an inspection. At the very least, explanation will be wanted if a valuable specimen is 'non-destructively' destroyed. Responsibility for the standard of an inspection is more subtle, and is usually covered by the contract. However, if the inspection is inadequately supervised, executed or reported, then difficulties are to be expected. The problems of reporting lead to a general rule: 'Don't make it hard for someone carrying out an inspection to report marginal defects.'

The inspector can be under pressure to suppress marginal cases for various reasons. If he or she is supervising construction in unpleasant working conditions in the field, with a workforce on piece rates, the inspector may feel physically threatened to accept doubtful welds. If the inspector is on an urgent project with tight schedules and penalty clauses, he or she may feel distinctly antisocial at causing a set-back by reporting a marginal defect. Some of these problems can be reduced by application of another rule: 'Keep inspection and manufacturing staff apart if possible'—though such good intentions may not be practicable.

Thirdly, in addition to strictly legal aspects, there are financial and moral components. The financial parts, i.e. the cost if a specific defect slips by and the economic effects of this, are quantifiable by routine methods. The moral parts bring in both general issues and more specific ones, like the effect on future trade if it were known that a large explosion occurred soon after your team had declared the system safe.

The main legal requirements relevant to this book include, for example, the validation of welded construction under the jurisdiction of the Factories Act 1961 and the Health and Safety at Work etc Act 1974 (see Barton 1982). Their precise interpretation, as for all Acts of Parliament, is a matter for a court of law. Nevertheless, they do introduce the notions of 'good construction', 'sound material', 'adequate strength', 'free from potent defects' and 'properly maintained', either as absolute obligations, or as obligations qualified by some level of practicability (see Barton 1982). In some cases, e.g. as regards construction, material and strength, it is generally considered that complying with the relevant British Standard suffices. Where there is no such standard, there may exist formal notes for guidance. So far as testing is concerned, similar standards, codes or formal notes exist for many areas. These include the ASME codes and British Standards (e.g. BS PD 6493 (1980)). Legal acceptance of specific approaches is inevitably a long-drawn-out affair. Clearly new developments in equipment and in testing may be delayed, so that formal notes, whilst of lesser status, are of some importance, and aim at a concensus of agreed good practice.

One consequence of ineffective inspection, whether from poor application, poor design or mere misfortune, is the risk of an undesirable event. We may divide these calamities into several types. First, there are the *expected* failures. One can be certain, if only from past performance, that bridges will fail and aircraft will crash. For these expected events, there will usually be statistics for probabilities, and there may be actions to reduce consequences, e.g. by providing longer prior warning. Secondly, there are *imaginable* failures (see, for example Marshall 1982), which are those which may never happen. The harbour bridge may fail at rush hour because it is struck by a crowded ferry which sinks; two large aircraft may collide and crash onto a full football stadium. Here there will be no statistics, and probability estimates depend on expert intuition, i.e. on guesses (as in §2.4.4). Between these classes come the areas where causal links are not certain. It is clear that testing, however good, can only make a modest contribution to the perceived risk of any but the expected failures. This leads directly to the subjective aspects of risk (§2.4.5). An inspection strategy which reduces death and injury may even be unpopular with a public which might prefer its fears assuaged by reinforcing its prejudices.

A valuable analysis of the cost of inspection is that of Groocock (1980) (see also Grant and Rogerson 1982). They define the *appraisal cost* as the cost of inspecting and testing items because of possible failure. This will include training costs, test design and the actual execution of inspection. Likewise, Groocock and Grant and Rogerson define the *failure cost* as the costs resulting from failure, including repair, reinspection, and loss of income. *Prevention cost* is that incurred in attempting to reduce total appraisal and failure costs.

This breakdown makes clear a balance between the cost of inspection and its benefits. This trade-off can be managed with efficiency when it is possible to match the inspection method accurately to the specific product design and needs. In many cases, contractural arrangements divorce responsibility for setting the validation requirements from the financial consequences of applying them. A good example is the well known committee factor of ten. If a critical level (e.g. of toxic compound) is x, as obtained by direct analysis of data, the critical level will have fallen to $(x \times 10^{-N})$ after being assessed by N committees, no further data being available at any stage. This separation of cost and need is both a waste of resources and a likely source of ineffective inspection, for it is only human to take less seriously procedures which seem irrelevant.

3

Defects and failure

When an error has been detected and corrected, it will be found to have been right in the first place

Non-destructive testing is used to find defects. Inspection techniques are neither able nor designed to detect all the different types of possible defect, but are restricted in their scope to certain specific types of defect. A first step in designing a reliable inspection technique is a working knowledge of the types of defect which might occur and the identification of those which are most likely to be detrimental to the component in operation. This chapter reviews types of defect, where and how they occur, and how important they are for different types of component. Both this and the following chapter on non-destructive testing techniques form a backdrop against which reliability of inspection must be set. For how can we measure reliability of inspection, or design reliable techniques, without some knowledge of the defects and the techniques currently used to detect and characterise them?

This chapter considers especially those defects which lead to failure and the mechanisms by which failure occurs. We consider therefore defects in raw materials, those introduced during manufacture, and those appearing in operation. The defects and defect types are discussed here in terms of the fracture mechanics framework, which we outline in §3.1. We shall note a number of places where the link with NDT has basic limitations. For example fitness for purpose describes a system as a whole, whereas NDT may monitor a single component only. Moreover, monitoring may have several meanings. The NDT examination may seek existing unacceptable faults, it may seek faults (like residual stress) which indicate propensity to become unacceptable, or it may involve comparisons at several times, whether by monitoring or fingerprinting.

Failure is not necessarily well defined. Generally it means a component is unacceptable for its job. Often, though not always, it means the item is unfit for its original purpose (e.g. an oil-stained raincoat may be cosmetically unacceptable, but may still suffice to keep out rain). For engineering materials, mechanical strength is the key factor, so that predictions of fracture, fatigue, corrosion etc, are involved. In medicine the detection of the unusual, e.g. tumours, and aid in diagnosis are the aim. For semiconductor devices the intention may be to avoid costly processing of faulty devices, or simply the estimate of how much redundancy is necessary. Our present concern is mainly with the engineering materials case, since it dominates NDT.

3.1 Fracture mechanics: mechanisms of failure

The ultimate aim of fracture mechanics is the control of fracture. In a given component, what mechanisms reduce effectiveness or endanger users, and how can their consequences be minimised? We turn the question around: what defects should we look for in NDT, and how should we recognise danger signs in the information we receive?

The main way in which fracture mechanics is used in practice involves a materials part and a geometric part. Both parts are designed to relate the behaviour of a real engineering component to standard laboratory tests. The materials part, for example, is concerned with the strength of a component of standard shape (perhaps a notched bar) as a function of chemical composition, microstructure, surface roughness, environment etc. The geometric part is concerned with the stress at the most vulnerable place for a specific shape of component under given external loading. Fracture mechanics provides the framework within which laboratory data and computed stress distributions can help to decide whether given defects are safe or not.

3.1.1 Modes of stressing

Failure under stress depends on the type of loading. Whilst real engineering components usually experience complicated combinations of several types of loading, it is useful to identify some special cases. Here we consider only the distribution of the load. Other factors, like time-dependent loads and environmental effects, will be mentioned as necessary.

First, there are three modes of plane and antiplane strain. These characterise the behaviour of thick plates, where any plastic region is much smaller than the plate thickness. This is the standard situation for ships, for bridges and for storage tanks. The three cases usually considered are illustrated in figure 3.1 and are

Mode I   : Plane strain resulting in cleavage
Mode II   : Plane strain resulting in shear rupture
Mode III : Antiplane strain, giving an alternative mode of shear.

Secondly, plane stress leads to tearing of thin plates, as of aircraft skins or

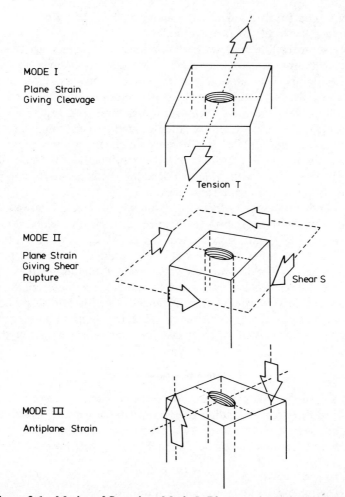

Figure 3.1   Modes of Stressing. Mode I: Plane strain under tension T, giving cleavage. Mode II: Plane strain under shear S, giving shear rupture. Mode III: Antiplane strain. Plane stress, e.g. in a thin sheet when there is no stress normal to the sheet, is not shown.

pipelines. In such cases there may be a plastic zone whose size is comparable with or even larger than the plate thickness. We emphasise that these three modes are idealised and real loadings may be much more complicated. In such cases computer-based finite-element methods come into their own.

3.1.2   Materials aspects: solids free from macroscopic defects

The mechanical properties of materials free from macroscopic defects are conveniently characterised by four parameters, three straightforwardly mechanical, the fourth, which is the melting point, being of more interest in the practical aspects of processing.

The most accessible and least ambiguous parameters are the *elastic moduli* (for the moment we may confine attention to the tensile modulus E and an analogous shear modulus) which measure how easy it is to deform a solid reversibly and elastically. The elastic moduli depend directly on the interatomic forces, and have a natural upper limit of a few times 10^{11} Pa (1 Pa = 1 N m^{-2}). At the other extreme, foamed polymers may have moduli of as little as around 10^{5} Pa. Just because an elastic modulus measures the response to small stresses so directly and reproducibly, a modulus like E is also a convenient unit with which to describe the other more complicated materials properties relevant for fracture.

The second parameter is the *yield strength*, σ_y, i.e. the stress at which permanent plastic deformation takes over from elastic behaviour. Again, this can vary widely, but rarely exceeds 0.1 E or 0.2 E, i.e. σ_y is a few times 10^{10} Pa (i.e. a few times 10^{6} psi). Generally ceramics have quite high yield strengths, whereas metals—notably very pure FCC metals—can yield easily.

These yield strengths are nominally for defect-free material, and so are only measured in very special circumstances, e.g. for fibres or silica rods. What is normally quoted is a calculated value, involving simple assumptions about cohesion in the case of cleavage, for which the yield strength (stress to cause failure) is

$$\sigma_f = (E\varepsilon_{coh}/b_0)^{\frac{1}{2}} \tag{3.1}$$

b_0 being an atomic dimension, and ε_{coh} is the cohesive energy of the crystal. For shear rupture corresponding approximations are made, this time about the sliding over each other of two sinusoidal planes, with interplanar spacing h and Burgers vector b. The yield strength is now:

$$\tau_m \sim (E \cdot b/2\pi h) \tag{3.2}$$

where, strictly, E should be the shear modulus.

The third parameter is the *toughness*, G, which measures the degree to which the solid can absorb maltreatment without breaking. G has the dimensions of stress, e.g. pascals or N m^{-2}, and a natural measure of relative toughness would be G/Ea, with a the interatomic spacing. Here one finds ceramics tend to have a low toughness, and fail easily after an energy input merely sufficient to compensate for the extra energy of the new surfaces. Metals, however, are often very tough because of the energy absorbed in plasticity. There is no inconsistency in having *low* yield strength (i.e. plastic deformation sets in readily) and *high* toughness (i.e. plastic deformation absorbs a large amount of energy prior to breaking).

Mechanical properties of engineering components clearly cannot be described by three parameters alone, invaluable though the elastic moduli yield strength and toughness are. Apart from mere complication from, for example, anisotropy or multiphase character, we mention three other important factors which add a time dependence to the materials properties. First there is *fatigue*. Repeated applications of a stress level which would not cause crack initiation or extension if applied statically can cause cracks to grow to critical sizes. This is known as fatigue. Fatigue is especially important for contact bearings and some aircraft structures. The fatigue resistance is an additional materials property. Note both the maximum stress S_r and minimum stress (or the maximum reverse-bend stress S_e for oscillatory loading) must be defined. Secondly, there is what is termed *static fatigue*. Materials may be weakened after a long period of stress loading. One mechanism here is hydrogen embrittlement, where hydrogen penetrates to key regions with the passage of time. Finally, we note the whole range of *environmental effects*, including stress corrosion cracking and corrosion fatigue, which all reduce the effectiveness of the material for its purpose.

3.1.3   Geometric aspects: defects and their effects

The important feature here is the extent to which defects cause stress concentration. Suppose we have a defect-free component under external stress. The stress (where, to avoid complicated notation, we discuss only one arbitrary component of stress) has values $\sigma_0(r)$ at various points throughout the component. When there is a defect present, under the same external loads the stress has values $\sigma_d(r)$. Clearly the places where $\sigma_d(r)$ is large are the ones where fracture is most likely to occur. These are generally the places close to crack tips. The degree to which the local stress is enhanced, $\sigma_d(r)/\sigma_0(r)$, is determined by a stress-enhancement factor, defined later. Such stress-enhancement factors contain most of the geometric information used in practice, e.g. the shape of the component, the position and orientation of the crack, and the nature of the applied stress. These factors are calculated by linear elasticity theory, which is only an approximation to real solids; nevertheless, such calculations are extensively tabulated (e.g. Sih *et al* 1977) and used for design rules.

   The standard example of stress enhancement is for a crack in the form of an infinitely long strip of elliptical cross section in an unbounded isotropic elastic solid (figure 3.2). If the cross section has a minor axis of essentially infinitesimal length and a major axis c (so that the total crack length in its cross section is $2c$) then the local stress near its tip (see figure 3.2 for notation) is given by:

$$\sigma_d(r, \theta) = \frac{K}{(2\pi r)^{1/2}} f(\theta) \tag{3.3a}$$

$$= \sigma_0 f(\theta) \, (c/2r)^{1/2} \tag{3.3b}$$

$$K = (\pi c)^{1/2} \sigma_0 \tag{3.4a}$$

$$= (\pi/2)^{1/2} \, (\text{crack length})^{1/2} \sigma_0. \tag{3.4b}$$

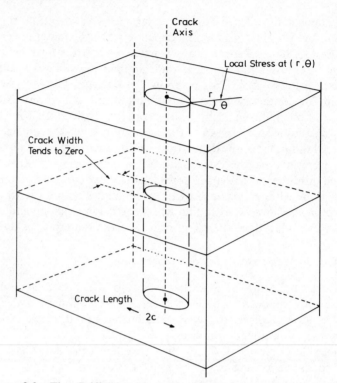

Figure 3.2 The Griffith crack, with elliptical cross section and of infinite extension in one direction. The 'length' is $2c$, and the coordinates relative to the crack tip (r, θ) are shown. Usually the crack is assumed to be very thin, i.e. its faces are essentially in contact when there is no opening stress.

The precise form of $f(\theta)$ depends on the mode of stressing, e.g. tensile or shear, and is listed in many places (e.g. Thompson 1983). The expressions given in equations (3.3) and (3.4) are, of course, subject to the limits set by describing this as the field *near* the tip, which implies $c \gg r$ and $r \gg \rho$, with ρ the curvature at the sharp ends of the elliptical cross section. Note the $r^{-1/2}$ dependence of σ_d: whilst the stress field is greatest close to the tip, it extends far from the crack tip too.

Clearly the local stress $\sigma_d(r, \theta)$ is enhanced over the nominal stress σ_0 by a factor

$$f(\theta) \, (c/2r)^{1/2} \tag{3.5}$$

which we may call the *stress concentration factor*. K itself is sometimes called the *stress intensity factor*, since the stress is given by the product of K and a purely geometrical component $f(\theta)/(2\pi r)^{1/2}$. When the maximum local stress σ_d is just sufficient to cause fracture, K becomes K_c the critical

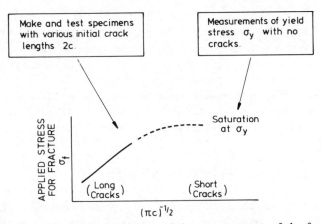

Figure 3.3 Schematic description of the measurement of the fracture toughness from measurements with various crack lengths $2c$.

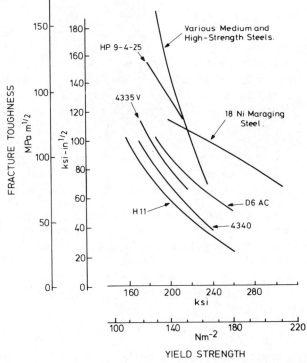

Figure 3.4 Fracture toughness and yield strength anti-correlate for medium-strength and high-strength steels. Data from various sources, including Pierce *et al* (1970) and Pellini (cited Guy 1971).

fracture toughness, which is a materials parameter for a given mode of stressing. Experimentally, K_c can be obtained by measuring the applied stress σ_f at which fracture occurs for several crack lengths. A plot of σ_f against $(\pi c)^{1/2}$ then yields an experimental value for this fracture toughness (see figure 3.3). It is an empirical fact that there is an anticorrelation between fracture toughness and yield strength, as noted in §3.1.2. Some examples are shown in figure 3.4. Designers have to trade off between resistance to plastic deformation and resistance to rupture.

3.1.4   Fracture of brittle solids: Griffith's theory

We begin with Griffith's classic example of a crack in a purely brittle solid. To be specific, we consider the case of plane strain, with purely tensile loading, though parallel calculations for other cases give very similar results. Griffith observed that the energy of a purely brittle solid containing a crack comprised two terms which depended on the crack length. First, there is the strain energy

$$\varepsilon_\sigma = - \sigma_0^2 \frac{(1 - \nu)\ \pi c^2}{2\mu} \tag{3.6}$$

whose sign indicates a reduction in energy as the crack is made longer and longer so as to relieve the strain. Here ν is Poisson's ratio and μ is a Lamé elastic constant. Secondly, there is the energy associated with the new surface created as the crack extends ($4c$ per unit length of crack, since there are two faces each of length $2c$), namely

$$\varepsilon_s = 4c\gamma \tag{3.7}$$

with γ the surface free energy. If extending the crack by c lowers the total energy, the crack will extend; if the energy does not fall, it will remain stable. With these expressions for the energies ε_σ, ε_s we find

(a) for a given external stress σ_0, all long cracks with full length $2c > 2c_{crit} = 8\mu\gamma/[\pi(1 - \nu)\sigma_0^2]$ will grow in length, and there is no bound, i.e. they continue to extend indefinitely. In some cases the critical flaw sizes can be very small. In the brittle ceramics like Si_3N_4 lengths as low as $10\ \mu m (10^{-3}\ cm)$ can be critical for applications to turbine blades.

(b) For a crack of given size c, there is a critical external stress σ_0^{crit} which leads to fracture:

$$\sigma_0^{crit} = \{4\mu\gamma/[(1 - \nu)\pi c]\}^{1/2}. \tag{3.8}$$

We can re-express these results in terms of a critical value of the stress intensity factor, K, and find a value:

$$K_{crit} = [4\mu\gamma/(1 - \nu)]^{1/2} \tag{3.9}$$

which depends on the materials properties μ, ν and γ alone. In Griffith's theory, fracture conditions have the important feature that they can be expressed in terms of separate geometrical and materials properties. In all such cases K_{crit} depends only on materials properties, but the precise functional form depends on geometric aspects. As a result, in principle at least, one can construct a list of values K_{crit} for various ways of loading, and it is values such as these which can be used to define acceptable crack lengths and loads. For example (Gordon 1978), for a large steel structure such as a ship or a bridge, a crack of length one metre might be acceptable up to stresses of 110 MN m^{-2}. For rubber, which can store a good deal of strain energy, the critical crack size even for modest loads would be only a fraction of a millimetre, which is why balloons burst with a bang when pricked by a pin. For brittle materials such as glass or ceramics the critical crack sizes, which limit the useful strength of these materials when mass-produced, are about one micrometre deep. Cottrell (1964a) quotes critical crack sizes of 25 μm in glass for a load of 2 MN m^{-2}.

3.1.5   Fracture of plastic and porous solids

Generalisations to plastic deformation and to inhomogeneous solids are usually intuitive. Thus the effect of plasticity might be presumed to change γ to $\gamma + \gamma_p$, where γ_p increases with fracture toughness. The plastic zone near the crack tip can be assumed to have a radius r_p related to the yield stress by

$$r_p \simeq (K/\sigma_y)^2$$

which, with the Griffith criterion, implies r_p values in the range of 0.3–1 μm and a strain of order 0.1% at the edge of this plastic zone. For our present purposes we shall not need the many more advanced approaches to the description of plasticity (see Thompson 1983).

In porous media, like cements, both the elastic moduli and the fracture surface energy depends on the volume fraction of porosity P. Kendall (1983) finds the elastic modulus varies roughly as $(1 - p)^N$, with $N \sim 2$ for closed-pore alumina and $N \sim 3$ for open-pore silica; the fracture surface energies vary more rapidly as $\exp(-dp)$, with $d \sim 2.2$ for a porous glass and $d \sim 7.7$ for a filled polymer. The effects of inclusions in steels are well documented (e.g. Wells and Hauser 1969); empirically, there is a modest fall in K_{crit} with inclusion content.

3.1.6   Relation of fracture mechanics and NDT

Since fracture mechanics includes the notion of a critical crack length for a given loading (c_{crit}), and since non-destructive testing measures crack sizes, it is tempting (but false) to assume the two are the same. Even for cracks of ideal geometric shape, there are limitations from the simple assumptions underlying linear elastic fracture mechanics and from the analysis of any chosen NDT method, and these simplifications may allow quite significant discrepancies. For real cracks, rough and oddly shaped, the differences can be greater still. There is a real need for experiments on controlled defects which relate specifically their actual fracture size and their actual NDT size. The importance of this limit is that it can reduce excessive and arbitrary safety margins, i.e. those common unscientifically based factors (which merely give a spurious illusion of safety) in favour of verified quantitative estimates of greater economy and assured validity.

The link between control of fracture and verification by NDT has other complexities which we note. First, changes in material may improve mechanical properties whilst (e.g. because of ultrasonic attenuation) making NDT more difficult. Improving a specification can make it harder to demonstrate that the specification has been achieved. Secondly, there can be more than one type of defect present, and one must be wary of choosing an NDT method in which a benign defect masks the problem defect. Thirdly, defect populations may evolve with time, whereas much NDT is done once only. The link between a current defect population in operation and that measured before use is a source of caution (sometimes over-caution, sometimes sensible reservation) which must be recognised.

3.2   Defects in starting materials: cast metals

Some defects can be detected before any significant processing has occurred. In steels, for instance, this means after casting and rolling, but before machining. For most purposes we shall concentrate on the more substantial defects and so be able to ignore grain boundaries, dislocations and other small-scale imperfections. This is not always the case, especially when one moves away from structural steels; problems of the quality of starting material are especially important when there is high added value, as for silicon integrated circuits. Nevertheless, the several types of cast metals show how complex defect structures can appear, and also how the microstructure may interfere with NDT.

3.2.1   Defects in continuously cast steels

Here (cf Mizoguchi *et al* 1981) we can identify three broad categories: zones of low ductility; cracks; and small volumes of different composition, e.g. inclusions, precipitates, pinholes, and their like.

Zones of low ductility.   The low ductility zones represent regions where there is enhanced segregation of particular impurities to grain boundaries, where special precipitates form, or where specific phase transformations occur to an undesirable degree. There is thus a distinct and characteristic dependence on the local temperatures achieved during the casting process.

(*a*) Zone I is the high temperature zone within 50 °C of the solidus temperature. Here the low strength arises because, in the dendritic growth of the solid from the melt, the liquid remaining between the dendrites becomes enriched in certain species. The enhanced levels of segregating elements (P, S, B) can lead to cracking in this region.

(*b*) Zone II occurs at intermediate temperatures, typically 900–1200 °C. Here the processes are complex. In austenitic steels, the reduction of strength can arise from the precipitation of (Fe/Mn) sulphides or (Fe/Mn) oxides in the grain boundaries.

(*c*) Zone III occurs at lower temperatures, usually 700–900 °C. Embrittlement is governed partly by $\gamma \rightarrow \alpha$ transitions of the steel, and partly encouraged by the intergranular precipitation of nitrides and carbides.

Cracks in cast steels. These are conveniently grouped into six classes, all with a distinctive origin; figure 3.5 shows their main characteristics.

(*a*) Centreline cracks. These develop from shrinkage cavities, and can be avoided by controlling the speed of casting.

(*b*) Halfway or internal cracks, sometimes known as radial cracks, result from tensile stresses in temperature Zone I; they can be controlled by adjustment of stresses and compositions.

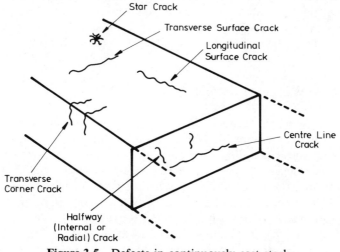

Figure 3.5   Defects in continuously cast steels.

(c) Longitudinal surface cracks result from non-uniform solidification and from shrinkage in $\delta \rightarrow \gamma$ transformations. These depend on the size of the cast slab and its composition, and can be limited by suitable design and mechanical properties of the mould.

(d) Star cracks arise from intergranular attack by Cu from the mould; Nb, V and Al may also cause star cracks, whereas Ni inhibits them.

(e) Transverse corner cracks appear to correspond to Zone III embrittlement during cooling, and can be controlled by heat treatment.

(f) Transverse surface cracks are probably caused by external stress, and normally occur along the valleys of ripple marks.

Regions of different composition. Non-metallic inclusions are commonly found, usually consisting of oxides, sulphides or silicates. They may affect workability or weldability. The larger oxide inclusions are often caused by fast cooling so that flotation is difficult to achieve. Aluminium inclusions and elongated inclusions of sulphides can cause problems if one wants a steel resistant to lamellar tearing.

Rolling of cast material alters the form of inclusions and hence their visibility in NDT. The non-metallic inclusions become laminations. Oxide and scale become scabs and bark, whereas metallic flaws lead to laps and slivers. Even gas bubbles are modified, becoming seams, The main effect is to produce thin, planar, defects.

Steels are, of course, not the only materials to be cast continuously. Another such material is float glass, and in this case it is interesting that the flaws are different. One example comes from tin drips. In the production of the float glass, molten glass is poured on to a molten tin bath where it is formed into a flat ribbon. High-quality float glass for mirrors can suffer from the flaws formed when vaporised tin condenses and falls on to the surface of the flat ribbon. After annealing, these tin drips appear as small cavities in the upper surface of the glass. These can then be detected by laser (Van der Neut 1981).

3.2.2   Defects in cast steel ingots

The cast metal which forms an ingot has properties which vary from place to place for two reasons. The first relates to the solidification structure, which is apparent metallographically. As one moves out from the centre (figure 3.6) one finds first equiaxed crystal growth, then a branched dendritic structure, a columnar dendritic structure, with finally a finely equiaxed chill zone at the outside. The second variation in properties relates to segregation and composition. Consider the fully deoxidised steel ingot of figure 3.6. Alloying elements are enriched at the centre near the top and deficient near the centre close to the base; in the lower, negative segregation region, silicates may be enhanced. Two types of 'ropes' of high inclusion concentration and positive segregation are found. The A-segregates in the dendritic regions result from entrapment of impurity-rich liquid, either between growing columnar crystals or behind equiaxed crystals nucleated ahead of the solid–liquid interface. These regions can be rich in MnS. The V-segregates, which occur closer to the central axis of the ingot, are similar but formed later in the solidification process.

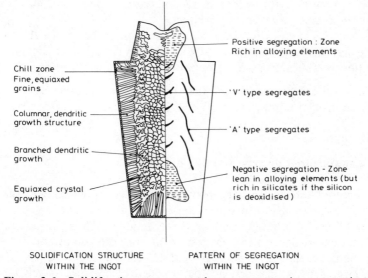

Chill zone
Fine, equiaxed
grains

Columnar, dendritic
growth structure

Branched dendritic
growth

Equiaxed crystal
growth

Positive segregation : Zone
Rich in alloying elements

'V' type segregates

'A' type segregates

Negative segregation - Zone
lean in alloying elements (but
rich in silicates if the silicon
is deoxidised)

SOLIDIFICATION STRUCTURE   PATTERN OF SEGREGATION
WITHIN THE INGOT   WITHIN THE INGOT

Figure 3.6 Solidification structure and macro-segregation pattern in a typical, fully deoxidised, large steel ingot (after Druce and Edwards 1980).

The complicated grain structure in cases like this, with its anisotropy and potential for ultrasonic attenuation, is an important factor in NDT, quite apart from any defects or apparent defects that may exist.

3.2.3   Defects in other cast metals

Comprehensive descriptions can be found in the *Metals Handbook* **5** (B) of the American Society for Metals, or in less detail in sources like Haywood (1952). The casting procedures affect the defects strongly, and the procedures reflect the different uses to which the metal is finally put. Casting can be continuous, giving steel strip, for instance, or it may be a moulding which is to be used after limited extra treatment.

Casting into ingots produces five main classes of defect, several corresponding to those described previously.

(*a*) Surface cracks, as previously, but also provoked by uneven cooling or by mould roughness.

(*b*) Slag inclusions, as previously.

(*c*) Blowholes. These arise from gas bubbles, either from dissolved gas being released on cooling or from oxidation of carbon. Blowholes near the surface can affect weldability.

(*d*) Cold shuts. These result from solidified splashes on the side of the mould. If they do not remelt on contact with the rest of the molten metal, one is left with a region whose surface may be oxidised and may contain segregates so as to weaken the product.

(*e*) Shrinkage holes or pipes. These occur near the top of the casting, and the surrounding metal is usually unsuitable for use as it is over-rich in impurities like phosphorus and sulphur.

Even this list is not comprehensive, for there are many ways in which perfect control of composition or heat treatment can be thwarted.

3.3 Manufactured defects: welding defects

Manufactured defects can arise in any of the stages of working of the basic material: cutting, surface preparation, heat treatment, coating, but above all in welding. In welding, cracking is the major problem, though residual stress and embrittlement in heat treatment are both important.

We shall concentrate on general weld defects, for this field provides one of the largest areas of practical NDT. Indeed, in pressure vessels made of good-quality steel, 90% of the defects will occur in welds (Marshall 1982). It also provides one of the areas of greatly improved standards and one where modern approaches to reliability are especially appropriate. As Rickover (1963) (cited in Burgess 1983) observed, the early standards for conventional components for nuclear submarines were seriously below acceptable. Even after long delays, substantial reworking was needed before specifications were met. One example of a set of carbon steel welds showed the following statistics for radiographic inspection to ASME standards:

Manufacturer's interpretation : 100% met specification
US Navy reanalysis   : 10% met specification
                               35% clearly failed
                               55% could not be interpreted
                                   unambiguously because of
                                   rough external surface.

One would like to hope the situation has improved.

The region between the unaffected base metal and the main bulk of the weld can be considered as three main layers (Savage (1978), and see figure 3.7).

The true heat-affected zone has altered microstructure, with grain growth, possibly resolution and precipitation of secondary phases. Mechanically, its behaviour will be reduced from the original base metal specification, and it is subject to stress-relief (reheat) cracking. The partly melted zone may well suffer hot cracking from microstructural changes initiated as microfissures. Cold cracking, especially hydrogen-induced cracking or intergranular cracking, may occur in the composite region adjacent to the unmixed zone. We now discuss these in more detail.

Here our discussion follows primarily that of the *Welding Handbook* of the American Welding Society and the review of Druce and Hudson (1982). Several of the defect types parallel those noted for castings.

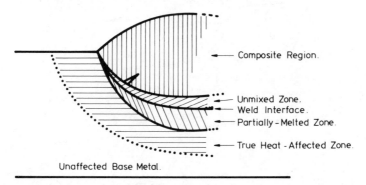

Figure 3.7 Structure of metallurgically distinguishable zones in a typical weld (after Savage 1978).

Porosity corresponds to gas pockets or voids in the weld, usually arising from chemical reactions during the welding. The reactions may involve contaminants in the joint, the flux, or the electrode.

Slag inclusions are non-metallic inclusions normally entrapped between weld beads, or between the base metal and the weld. These inclusions, often oxides, are insoluble in the molten weld metal and, being less dense, rise to the surface where they are entrapped.

Tungsten inclusions. Here particles from the welding electrode become embedded.

Cold cracking occurs when the steel is insufficiently ductile. It may involve hydrogen, or it can be associated with stress concentration at notches or inclusions.

Hot cracking (or liquation cracking) is mainly associated with low melting point constituents in the weld metal. These, e.g. Fe sulphides in steel, accumulate along grain boundaries during solidification to give weakened zones. Intergranular or interdendritic cracks form as the liquid cools. The region cracking is usually rich in elements like S, Sn, Cu, P or As. Crack sizes typically range from a few micrometres to a few millimetres. Their preferred orientation in welds (there is no preference in forgings) is shown in figure 3.8(*b*).

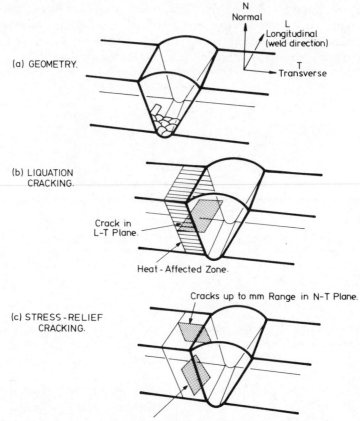

Figure 3.8 Orientations of cracks in welds: (*a*) definitions; (*b*) liquation cracking; (*c*) stress-relief cracking.

Reheat (stress-relief) cracking results during the heat treatment to relieve residual stress after welding or some similar processing. The cracking occurs when the relaxation strains exceed the local ductility of the material. It occurs in the heat-affected zone, and not in the weld metal. The major cracks can be large—up to several hundred millimetres are reported—and lie parallel to the fusion boundary (figure 3.8(c)). Microcracks, sometimes associated with macrocracking, tend to lie normal to the welding direction.

Incomplete fusion refers to a failure to fuse together adjacent weld regions or weld/base regions.

Undercut refers to the consequences of melting the sidewall of a groove, for example, so that there is a sharp recess just where the next bead is to be deposited. Undercut also describes the reduction in thickness at the line where the beads in the final layer of the weld metal tie into the surface of the base metal, for example at the toe of the weld.

Other defect types include *lack of penetration*, *overlap*, and *concavity*.

It is an impossible task to give a comprehensive picture of the frequency with which weld defects occur. However, we can quote (table 3.1) figures listed by Rogerson (1983) for the number of indications recorded exceeding 2 mm in each metre of weld listed.

Table 3.1 Indications larger than 2 mm per metre of weld.

	NDT method†	Submerged arc welded	Manual metal arc welded	Gas–metal arc welded
Low-alloy steel pressure vessel	X	0.3–0.5	0.7–1.0	
Steel tankage, site welded	X	3.4	0.7–4.9	
Aluminium pressure vessel				1.6
Ship's hulls	X		4.9	
Nodes in tubular offshore platform jacket	A		5.9	
Typical problems for named weld process		*Porosity* due to fast cooling *Hot cracking* of thick plates *Grain growth* as high heat input	*Porosity* due to absorbed nitrogen from air *Residual slag* inclusions	*Porosity* due to CO_2 gas absorption

† X = radiography, A = ultrasonic.

3.4   Defects which grow in operation

Many systems show a characteristic time-dependent failure rate. For mass-produced items this often has the so-called 'bathtub' form (figure 3.9). At short times there are failures of freaks in the population—the 'infant mortality' of the items which slipped past inspection. The number of freaks usually reduces as manufacturers learn tricks and tests to avoid their causes. For intermediate times, there is a low rate of random failure. Finally, components begin to wear out, and the failure rate rises steadily. In the present section, we are mainly concerned with the stage which would be labelled 'wearout'. At least one key defect has grown to an unacceptable size.

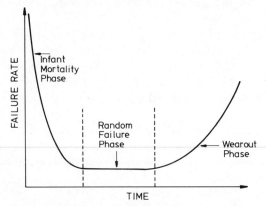

Figure 3.9   The 'bathtub': the dependence of failure rate on the age of the component.

The 'wearout' phase often shows characteristic features. If there is only a single mode of failure, one often finds a lognormal distribution of lifetimes, i.e.

$$\log (\text{lifetime } \tau) = A + B (\text{fraction } f(\tau) \text{ failing up to time } \tau)$$

where A and B are constants for a given system. The parameters A and B are usually determined by *accelerated testing*. A common form is to look at data for higher temperatures, for many failure processes show the so-called Arrhenius behaviour, with the median lifetime proportional to $\exp(-C/T)$, with C a temperature-independent parameter.

Other damage which can result from a cumulative effect in operation may be of quite a different type. Electrical breakdown is an example, where there is a threshold of some sort in the electric field (whether optical, as in laser damage, or the static field across a capacitor; see Hayes and Stoneham (1985)). Here a gradual degradation can occur, notably through field-induced changes in microstructure. In the human body, cumulative heavy metal poisons are analogous.

3.4.1   Fatigue

Fatigue, in which repeated loading causes failure under smaller loads than under static loading, has been studied systematically since the mid nineteenth century. The concept of fatigue leads to problems. If a specimen fails in a way which depends on its stress history, with what parameters can one quantify the fatigue resistance of a material? How can the time-dependent stress be characterised when it is not simply periodic?

The working rules adopted (and they are more a classification of data than demonstrated scientific laws) are the following.

The Paris equation. Suppose the applied stress is such that the stress-intensity factor oscillates with amplitude ΔK. The rule asserts that the crack extension, a, varies with number of cycles such that

$$\frac{\mathrm{d}a}{\mathrm{d}N} = D\,(\Delta K)^m$$

where D and m are materials parameters; typically $m \sim 3$. The result needs some qualifications, namely

(*a*) There is dependence on $R \equiv K_{max}/K_{min}$, i.e. on the size of any static stress component. This dependence may be modest, as for mild steel, or strong, as for high-strength Al alloys. The static stress can be a load, or may be merely residual stress.

(*b*) There is a threshold at low ΔK corresponding to crack extensions of about one atomic spacing per cycle.

(*c*) There is a stage at which crack extension becomes very rapid as K_{max} approaches the yield stress.

(*d*) Whilst the equation implies smooth continuous extension, any inclusions may lead to steps in crack extension. Such steps are one contribution to the spread of values of plots of $\mathrm{d}a/\mathrm{d}N$ versus ΔK. Since the steps correspond to a different mechanism of extension (i.e. they do not represent stable growth through the matrix), they can suggest misleadingly conservative expressions for the fatigue growth parameters D and m.

(*e*) Note there is no explicit dependence on frequency. The time to failure (which can be obtained by integrating the equation) is simply the product of period per cycle and number of cycles failure.

Miner's rule (the Palmgren–Miner equation, Palmgren (1924), Miner (1945)). When several stress levels are involved, if there are n_i cycles at stress level i, and if N_i cycles at that level would cause failure, the damage is presumed to accumulate linearly until a critical value is reached. Thus failure occurs when:

$$\sum_i n_i/N_i = 1$$

Numerous generalisations exist.

For probabilistic analysis, what is needed is the probability $P(a \mid a_0, \Delta K, N)$ that a fatigue crack has length a, given its initial length a_0 and N cycles in which K varied by ΔK. This probability reflects statistical factors, such as the presence of inclusions. One effective representation is the following (Temple 1985a):

$$P(a \mid a_0, \Delta K, N) = \frac{1}{\sqrt{2\pi}} \frac{1}{\sigma(\Delta K)} \exp \left[-\frac{1}{2} \left(\frac{a - a_0 + Ng(\Delta K)}{\sigma(\Delta K)} \right)^2 \right]$$

where $g(x)$ and $\sigma(x)$ are fitted to forms linear or quadratic in x. The main disadvantage of a normal distribution is that it will predict negative values of crack growth, albeit with very small probability. This can be overcome by truncating the normal distribution at zero and renormalising. Taking account of the variation of crack growth rate with R gave two additional equations for the mean growth per cycle, in millimetres (Temple 1985a)

$$g(\Delta K) = -4.70 \times 10^{-4} + 7.30 \times 10^{-5} \Delta K$$

for $R < 0.6$ whilst for $R \geqslant 0.6$ the best fit obtained is:

$$g(\Delta K) = -1.02 \times 10^{-3} + 1.46 \times 10^{-4} \Delta K.$$

The standard deviations that go with these means are given by:

$$\sigma(\Delta K) = 4.23 \times 10^{-5} + 2.72 \times 10^{-5} \Delta K$$

for $R < 0.6$ whilst for $R \geqslant 0.6$ the best fit obtained is

$$\sigma(\Delta K) = -2.95 \times 10^{-4} + 6.21 \times 10^{-5} \Delta K$$

for stress intensity increments ΔK in $MPa\,m^{1/2}$. These expressions give a threshold value of ΔK. Experimentally a threshold value of ΔK is found (e.g. Scott et al 1983) which is typically about the same magnitude as this analysis reveals, that is about $7\,MPa\,m^{1/2}$. However, in practice the threshold value decreases with increasing R-value whereas that predicted by the equations above increases slightly with R-value. This suggests that the particular threshold values found here are probably artefacts of the particular database used in the analysis. Now consider what happens as the cracks grow in a reactor. In this environment there are a number of transient stresses (tabulations can be found, for example in Harris (1977) and Marshall (1982)). In most probabilistic assessments of vessel integrity some assumptions are made regarding the type of crack and its location in the vessel wall. These assumptions are necessary in order to translate the applied stresses into stress intensities at the cracks. This stress intensity is then compared with some fracture criterion, such as is given by linear elastic fracture mechanics,

$$K_I \geqslant K_{IC}$$

in order to test whether crack extension initiates. Typical assumptions made about the cracks are that they are extended line cracks, or surface breaking semi-elliptical cracks with a given aspect ratio. Suppose we want to calculate the growth that occurs after n_1 cycles which produce stress intensity increments ΔK_1 and n_2 cycles which produce ΔK_2 on our chosen crack. The new mean crack size a will be given by:

$$a = a_0 + n_1 g_1(\Delta K) + n_2 g_2(\Delta K)$$

where a_0 is the original crack size. For convenience in the probabilistic calculations, the final size is taken to be independent of the order in which the cycles are applied. In reactor operations the cycles of different stress levels are assumed to be spread uniformly over a 40-year life which will tend to average out any history effects.

For the standard deviation σ about this mean value we will obtain:

$$\sigma^2 = [n_1^2 \sigma_1^2 + n_2^2 \sigma_2^2].$$

3.4.2 Stress-corrosion cracking

Stress-corrosion cracking is one of a variety of ways in which the environment of a component (e.g. sea-water) can make it much more sensitive to applied stresses. The physical and chemical processes are many and varied, ranging from the stress-induced cracking of protective oxide to allow access to an aggressive environment to the build-up of especially high concentrations of hostile species in mechanically opened pits. Among the rules describing stress-corrosion cracking in many cases are these:

(a) The time to failure depends on some power of the applied stress, which can be written

$$\log \text{(time to failure)} = A + B \log \text{(stress)}$$

with A and B time dependent.

(b) There is a stress-intensity factor K_{scc}, less than K_{crit} of equation (3.9), such that cracks will grow in the presence of a hostile environment. The crack grows more readily and rapidly the larger the value of K, of course.

Clearly one way to control stress-corrosion cracking is to ensure initial crack sizes are kept below an even smaller critical value than those for stress alone. In stress-corrosion cracking there is also the possibility of monitoring, i.e. exploiting the electrochemical nature of the process to measure a corrosion current. Whilst this is not always possible, it does represent another approach to non-destructive examination.

3.4.3   High strain rate defects

Under dynamic loading, whether by missile impact or geophysical event, the rate of loading becomes critical (Curran *et al* 1977). Dislocation velocities and crack nucleation rates become important. The various models fall into two main groups.

(*a*) Passive models, in which the degree of materials failure depends on the prior stress history, but it is assumed that the dynamical loading in a particular test does not affect the stress history appropriate for that test. The Tuler–Butcher model (Tuler and Butcher 1968), for example, suggests a level of damage given by

$$D = A \int_0^t \mathrm{d}t \; A(\sigma(t) - \sigma_0)^\lambda$$

where A, λ and σ_0 are materials constants. The Zhurkov model (Zhurkov 1975) is essentially a creep-rupture model, and indicates a time of failure

$$\tau = \tau_0 \exp \; [(U_0 - \gamma\sigma)kT]$$

where τ_0, γ and U_0 are materials parameters.

(*b*) Active models, where there is a continuous modification of the materials parameters during the dynamical loading. Nucleation and growth models are of this type. If $(\mathrm{d}N/\mathrm{d}t)_0$ is a threshold nucleation rate and the two characteristic stresses are a tensile threshold (σ_0) and a stress sensitivity (σ_1) for nucleation then a characteristic relation is:

$$\frac{\mathrm{d}N}{\mathrm{d}t} = \left(\frac{\mathrm{d}N}{\mathrm{d}t}\right)_0 \exp[(\sigma - \sigma_0)/\sigma_1].$$

In this case the average growth velocity for microcracks also has a threshold, and is bounded by the Rayleigh wave velocity.

The links between these expressions and non-destructive testing methods take several forms: through microcrack sizes for very brittle solids, through residual stress measurements (in relation to how readily a threshold stress is achieved) and in materials characterisation in key regions of the specimen.

3.4.4   Ceramics

Ceramics have their own special problems in NDT. Since they are usually brittle—and often close to ideally so—one might imagine Griffith's theory would cover all important issues. The special problems reflect two special factors:

(*a*) The critical flaw sizes for many applications are very small, e.g. a few tens of micrometres. In polycrystalline ceramics (e.g. sintered or hot-pressed powders) the critical flaw size may be comparable with the grain size. Moreover, machining flaws can have very great importance. Even with high-quality ceramics and careful machining, corner chipping and other machining damage can dominate (Lenoe 1979).

(b) There can be a lack of reproducibility because nominally identical specimens differ in detailed microstructure. This has consequences for both mechanical and other (non-mechanical) performance standards. For example, with oxides for substrates of electronic devices the alignment of different regions is critical, so that the shrinkage during heat treatment can be controlled accurately. The degree of alignment of the anisotropic alumina grains proves to be of real importance; fortunately it can be monitored non-destructively by spin resonance (Mehran et al 1981). This example shows the problems of characterisation and the varied forms it can take.

When fracture is critical (as for optical fibres (Tariyal and Kalish 1977) as well as for engineering ceramics), statistical methods, especially those based on Weibull statistics (see the Appendix), have proved especially important. Weibull statistics do not always suffice, however, nor do the multimode forms proposed as replacement necessarily have a secure scientific basis. For example, in the Appendix we cite data showing that the statistics of glass fibre fracture depend on the mode of preparation, i.e. Weibull statistics are not the only form for a single class of system. Time-dependent extensions of statistical approaches have been discussed by Davidge (1979).

The flaws in engineering ceramics can be conveniently divided into groups (Bortz 1979): *intrinsic microstructural features* such as large grains, pores, inclusions, impurity-weakened grain boundaries, etc; and *extrinsic surface defects* from surface finishing or from damage incurred in service. Note that the two types of defect may be interdependent, and that the defect population may change during service, i.e. predictions from statistics reflecting only the *initial* defects may mislead. These defects themselves may not weaken the structure in their initial form. Their importance is that they initiate, encourage or grow into harmful cracks.

If the only method of strength degradation of ceramics is subcritical crack growth from pre-existing flaws, then there is no difficulty (see e.g. Rutter et al 1979) in predicting the evolution of the crack distribution (albeit subject to simplifying working approximations). Likewise, the inverse problem can be answered, namely what initial flaw distributions will give only acceptable cracks in the assumed lifetime?

There are, inevitably, many complications. Environment dependence is one: fatigue failure of alumina, for example, is influenced by moisture as well as temperature. SiC and Si_3N_4 in oxidising atmospheres can show both enhanced subcritical crack growth because of enhanced plasticity and thus more rapid creep fracture, and also crack healing by oxide formation.

Indeed, as Rutter *et al* discuss, it is the combination of flaw healing, pit formation and subcritical crack growth which determines lifetime in ceramics. Clearly, the lessons for advanced applications are first that there should be an adequate definition of working environmental conditions for the component, and secondly that there should be a sufficient scientific understanding of the processes involved; mere data extrapolation can be unreliable. Table 3.2 summarises the present status of NDT for a representative ceramic, silicon nitride.

Table 3.2 Non-destructive examination of silicon nitride.

Method	Defect type, size	Comment
Surface and near-surface defects		
Dye penetrant	~ 250 μm	Limited sensitivity. Influenced by surface conditions.
Scanning laser acoustic microscope	~ 100 μm	Still at research stage. Subsurface laminations can be detected. Influenced by material porosity.
Laser photoacoustic spectroscopy	~ 50–100 μm at surface ~ 100 μm near surface	Still at research stage.
High frequency (250 MHz) ultrasonics	~ 25 μm	Contacting method. Still at research stage.
Bulk defects		
Microfocus x-ray with image enhancement	25–250 μm inclusions depending on density	Only modest sensitivity. Thicker samples (⩾ 0.6 cm)
Ultrasonics (45 MHz)	50–125 μm inclusions independent of density	Depends on geometry of component; influenced by attenuation for reaction-bonded Si_3N_4.
Ultrasonics (250 MHz)	25 μm inclusions	Still at research stage.

3.5 Defects in non-engineering systems

Even though the present book is concerned almost entirely with structural integrity and load-bearing engineered systems, the problem of reliable inspection is far wider. We cannot hope to cover all the broader issues, but we do feel it useful to mention some other situations. These have different criteria for success, different constraints and different technical capabilities. Inspection for defects which affect structural integrity can only benefit from being seen in a wider perspective, especially with the demands from moves towards new classes of engineering materials and new structures.

3.5.1 Biological systems and medical NDT

Here we find situations far removed from traditional engineering NDT, yet with parallels. Perhaps non-destructive *diagnosis* is a better term. The *non-destructive* nature of the test is obviously critical in medicine, where the patient (or their surviving relatives) can object in strong terms. Even mild probing may damage a sensitive foetus or internal organ, so special care is essential. There is, however, the advantage that one can often speak to the patient; very few specimens can be so easily interrogated.

The main aims of medical NDT are of two types. First, one may wish to look at the sizes and shapes of organs (including the foetus). Secondly, one may wish to measure the state of the body fluids, e.g. cell counts, or air bubble concentrations in testing for bends in divers. Several different methods of NDT are used. X-rays are common and widely known. They have disadvantages, e.g. the discomfort of the barium meal, the significant radiation level, and the relatively poor resolution (though this has improved enormously with use of new phosphors and digital recording methods). Nuclear magnetic resonance offers the monitoring of individual molecular species, albeit with low spatial resolution. Radio tracers require the body to do any concentrating necessary, and so merely highlight critical regions.

Ultrasonic methods have advantages, e.g. convenience and safety, provided really high resolution is not needed and provided attenuation is not so high as to need too high intensities. One problem is the variability of velocity and attenuation. Whilst sound velocities range from about 2460 m s^{-1} in fat to around 1600 m s^{-1} in muscle, values for nerve tissue, for instance, range from $1500-1600 \text{ m s}^{-1}$. Attenuation (usually expressed as the ratio of the attenuation α, where the intensity falls off as $\exp(-2\alpha x)$ with distance, to the angular frequency, f) also varies substantially from around $0.1 \text{ dB cm}^{-1} \text{MHz}^{-1}$ for blood to $20 \text{ dB cm}^{-1} \text{MHz}^{-1}$ for lung or skull bone. The mechanisms of attenuation in body fluids are themselves sometimes different from those in solids, and include (Woodcock 1979) the obvious viscous absorption and thermal relaxation, plus the consequences of blood being a suspension in which the suspended particles do not follow completely the fluid motion (see Atkinson and Berry 1974, Carstensen and Schwan 1959).

3.5.2 Semiconductor systems

Here non-destructive testing follows a completely different strategy. Whereas for large fabricated engineering structures testing follows the strategy:

(i)   build the structure (or part of it);
(ii) test non-destructively;
(iii) either accept or repair as necessary;
(iv) if required, retest until acceptable.

For semiconductor systems the approaches follow the strategy:

(i)   build a batch of many samples (e.g. devices);
(ii) test a few destructively;
(iii) either accept or reject the whole batch;
(iv) if rejected, adjust the manufacturing process.

Any one sample is either untested or destroyed (note that mere removal of encapsulation amounts to destruction of a device).

There are, of course, other more conventional strategies, but these are adopted where the high added value of fabrication is important. In ceramic substrates, for example, two of the requirements may be low natural radioactivity, to prevent hard and soft error generation in use, and imprecise compaction leading to misalignment in multilayer devices because of a degree of grain alignment outside given criteria. Here special methods have been developed, in these cases using autoradiography and spin resonance (Mehran *et al* 1981). Other fairly conventional NDT is done on silicon wafers, where optical microscopy is used for surface dislocations, ellipsometry for wafer concavity, and electrical measurements for other diagnoses.

Nevertheless, the testing of semiconductor systems relies principally on the statistics of sampling and the use of destructive tests on samples to establish modes of failure (e.g. whether open or short circuit), mechanisms of failure (e.g. corrosion or electromigration) and causes of failure (e.g. poor design, overloading etc). Failure can result from interaction between components, a notable example being that the packaging of a device (i.e. its mounting on ceramic substrates etc) can introduce stresses up to 60% of the breaking stress of silicon; even if fracture does not occur, device characteristics can alter. The various physical and analytic methods are reviewed by Richards and Footner (1983) and Stojadinovic and Ristic (1983); other aspects are surveyed by Buehler (1983) and Bertram (1983).

3.5.3   Failure of semiconductor devices

Semiconductor devices are composite systems which show time-dependent failure in their several different constituent parts. Oxide layers and their interfaces may degrade with charge accumulation or from dielectric breakdown; the metallisation may fail from electromigration or galvanic corrosion; bonds, and indeed any mechanical interfaces, can fail by fatigue or by intermetallic growth. There can be more generalised degradation, e.g. failure of seals leading to loss of hermeticity. Some of these problems will become more acute with trends towards smaller and smaller devices.

Accelerated testing procedures are standard for new devices. These take forms such as operation at normal voltages and currents, but at higher temperatures (possibly cycling the temperature) and with varied humidity.

We note now some representative results. First, failure versus *time* shows the so-called 'bathtub' curve (figure 3.9) mentioned earlier: a high 'infant mortality' as the freak members of the device population fail at short times, a period of random failures with low failure rate, and a gradual rise in rate as devices wear out at long times. Secondly, if there is a single dominant failure mode, a lognormal distribution of (cumulative) failures versus lifetime holds. Thirdly, many failure mechanisms show an Arrhenius dependence on temperature, i.e. the failure rate varies as $\exp(-\theta/T)$. The characteristic temperature θ can be obtained from measurements at several temperatures. However, there can be a vast difference between lifetime 'on the shelf' at a given temperature and the lifetime in operation. This is especially true of semiconductor lasers and other devices with high carrier densities, where (as noted earlier) recombination-enhanced degradation (see Stoneham 1981) is important. A fourth point is clearly that several mechanisms may operate, each with their own regimes of importance. The simple lognormal distribution may need generalisation. Finally, the relative importance of failure modes can vary when one merely changes the size of a device. The physical processes scale in different ways with dimensions. Once more, there is a real need for scientific understanding in technological prediction.

3.5.4   Laser damage due to pre-existing flaws

Pre-existing flaws can lead to both loss of performance and destruction in optical systems as well as in the better-known examples from mechanical engineered structures. Thus in these systems, like solid state lasers, there is an opportunity for non-destructive testing to check initial materials and final devices for their propensity to fail.

One common mechanism of failure initiated at flaws is mechanical failure (e.g. the cracking of the glass of a laser) because of stresses generated by local thermal expansion resulting from strong optical absorption of light at the flaw. Another is the result of focusing of light by the flaw, so that the local electric field exceeds the breakdown field; this is distinct from self-focusing damage, which involves the intrinsic non-linear refractive index rather than a flaw.

Even if mechanical damage is avoided, efficiency may fall because of defect growth, notability through recombination-enhanced dislocation climb. In all these mechanisms, there exist non-destructive optical methods for assessing material or devices. These specialised methods of non-destructive testing will surely become increasingly important as optical systems are used both by themselves and in monitoring mechanical systems.

4

Methods of non-destructive testing

The law of non-destructive thermodynamics: things get worse under pressure

4.1 Introduction

We now turn to the techniques of non-destructive testing. Our brief survey is intended partly to act as a preliminary to the discussion in Chapter 5 of the failure of inspection, and partly to give some idea of the problems and compromises in any practical approach to testing. These compromises can be of a very general nature, e.g. rapid inspection of a large volume is normally inconsistent with very precise locations of defects (a point which affects the strategy of inspection), or, in both ultrasonics and in eddy-current techniques, one finds a trade-off between resolution and the region of the specimen accessible because of attenuation. Thus the present chapter will outline the physical principles of the several techniques and draw attention to some of the likely sources of error. These aspects identify the points of concern for reliability which are analysed later.

We shall group the many techniques into four broad categories. First (§4.2) there are direct visual methods. Secondly, we cover the several approaches which are broadly 'acoustic' (§§4.3–4.6), in that they involve propagation of elastic vibrations within specimens, however generated and monitored. Thirdly (§§4.7–4.9), we cover the techniques which exploit electromagnetic methods. Next, we describe the different radiographic approaches in §4.10. Finally, in §4.11 the concept of proof testing is discussed. Clearly it would be impractical and inappropriate to survey all these techniques in detail, and we refer to the specialist literature for such a description (Fordham 1968, Krautkramer and Krautkramer 1977, McMaster 1982, etc). Our aim is to identify the physical principles and hence the likely sources of error. Knowledge of the nature of the relative strengths and weaknesses provides a basis for a fuller discussion of reliability in Chapter 5.

4.2   Visual inspection

This is the most basic approach to NDT. Perhaps, in consequence, it is probably the most disregarded. It is easy to underrate the importance of visual methods, yet they protect one against some really gross errors, and their omission has probably contributed to serious failures; both the 1879 Tay Bridge and the 1980 Alexander Kielland oil rig disasters appear to illustrate the need for visual methods.

In the case of the Tay Bridge the inspections used in the foundry were inadequate. The inspector used merely the sound of a hammer blow on a column of the bridge to determine its thickness, whereas it would have been a simple job to measure the thickness with a pair of calipers. In practice the inspector could not distinguish between columns which were only about 16 mm thick compared with a design thickness of about 25 mm, even though this difference would have been obvious visually, without any measurement at all. Nor was construction of the bridge supervised appropriately, and defects were apparent before the disaster occurred. The Alexander Kielland was a semi-submersible rig in the North Sea which capsized on 27 March 1980 with a loss of 123 of the 212 people aboard. The apparent cause of the disaster was a crack in a 6 mm filet weld for a sonar flange.

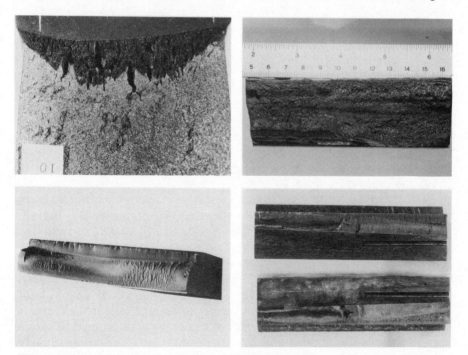

Figure 4.1   Top left, ductile cracking; top right, crack-like welding defect; bottom left, fatigue crack; bottom right, lack of side-wall fusion.

The crack propagated by fatigue in the cyclical stresses, imposed by waves, tides and the wind, until one of the five legs (each 8.5 m diameter) broke. The crack could have been detected visually before this occurred, but was not inspected, for the particular member had not been designed to contain any stress concentrations. The sonar flange turned out to be just such a stress concentrator.

Whilst limited to surface defects or surface condition, visual inspection can nevertheless identify potential sources of trouble at an early stage. It should form a part of the most sophisticated inspection systems, not just the most primitive. Indeed, visual inspection is used extensively in the examination of nuclear reactors, in this case using remote TV systems. An important recent example of the need for this component of NDT was given in the Seminar on the DDT trials by Rogerson *et al* (1984): 'We took the precaution of looking at the plate and saw a defect.' This shallow defect was, in fact, missed by many of the inspection teams taking part.

Visual inspection covers a wide field, ranging from sophisticated TV monitoring systems to the simple action of picking up a specimen in the hand. In addition, aids to visual defect location, such as the use of magnetic inks and penetrant dyes, provide important fields of study which are inspection techniques in their own right. Figure 4.1 shows a variety of defect morphologies.

4.2.1   Visual inspection and monitoring

Whether a TV system, microscope or the unaided eye is used it will be clear that a visual examination of the surface of a specimen can both locate defects and give warning of changes in general condition. In need not be necessary to touch or handle the specimen, for good-quality TV systems can be used for inspection in very hostile environments. For such purely optical inspection, there are no definite formal rules set down regarding what to look for, unless clumped all together under the global term 'changes'. Cracks may be directly visible, visible through cracking or distortion of a surface layer or shown up by the presence of corrosion products. There is necessarily some degree of art in the interpretation of the field of view, and there is no doubt that individuals vary in their ability to notice these effects. On the other hand even experienced workers can be fooled by the presence of scratches, machining marks etc, into 'seeing' defects which are not there. In the case of visual inspection, therefore, there is no clear definition of 'reliability', and we see the human factors in an especially acute form.

In addition to locating defects, visual techniques can be used to detect generalised changes in the system as a whole, due perhaps to wrong running conditions etc. Such changes may be ill-defined, though clearly present, and their recognition may prevent the build-up of more serious trouble later.

4.2.2   Aids to visual inspection

Simple visual identification of defects, particularly cracks, can be 'hit or miss'. There is an incentive to make the potentially serious, surface-opening, crack stand out more readily and consistently. Two techniques are in common use to achieve this. The first, discussed by Betz (1963), is the use of dye penetrants, which can be used to stain the crack line. The second is the magnetic particle inspection technique (Lewis 1951), in which the stray field attracts suitably dyed magnetic 'ink' to the crack region. These techniques are discussed below. They will be more effective in enhancing crack-like defects than any volumetric defects (pores, blow holes, cavities, etc). In qualitative terms, this is summarised in table 4.1, and shows the defect-specificity of reliability.

Table 4.1

Defect type	Visibility	
	Visual alone	After dye penetrant or magnetic particle enhancement
Crack-like	Poor	Good
Volumetric	Moderate	Moderate

Dye penetrant enhancement

Certain liquids can penetrate into the space between two surfaces separated by only a narrow gap, such as occurs in 'tight' cracks. They will also enter cracks which are open to the surface. If these liquids are applied to the specimen surface, the result is that the crack becomes filled or partially filled with the liquid. So if the liquid is coloured, or if it carries a brightly coloured dye, once the surplus liquid has been removed, the crack opening should stand out more clearly than when the crack was in its original state (figure 4.2). An alternative approach with the same physical basis is to use a liquid which fluoresces under ultraviolet light, so that the crack will become clearly visible when examined under ultraviolet radiation. This is especially useful in giving a clearer picture when TV systems are in use.

The main disadvantage of the dye penetrant technique is that it relies on the properties of the penetrant liquid to enter the crack, and this will be affected by the crack condition. Thus cracks already filled with liquid or corrosion product, or cracks which are very tight, may not allow the ingress of the penetrant, and so may not show up. On the other hand, shallow surface features, such as scratches, may allow some ingress of penetrant, and these would show up on the record as clearly as more important defects.

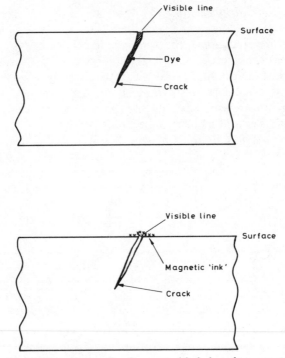

Figure 4.2 Aids to visualisation provided by dye penetrant and magnetic particle inspection techniques.

Application of the technique must allow for these possibilities, raising again the issue of judgment in reliable NDT by visual methods.

Magnetic particle inspection

If a tangential magnetic field is applied to the surface regions of a ferromagnetic specimen, it will normally lie totally within the specimen. However, if the specimen surface is cracked, a portion of the field is forced to leave the specimen locally, forming a stray field on the surface of the specimen. Magnetic particles will be preferentially attracted to these regions of stray field. The magnetic particle technique relies on the application of a magnetic 'ink', containing many minute magnetic particles which will tend to become attracted to any regions of stray field. Assuming that the ink is clearly visible, aided possibly by the addition of a dye, the regions which are defective will be clearly delineated (figure 4.2).

The technique has the advantage that the ink is positively attracted by the stray field, and that the magnetic field is affected by subsurface defects too, although the magnitude of the stray field generated falls off rapidly if the defect does not break the surface. On the other hand the technique is limited to application on ferromagnetic materials, which is a drawback, even though this class of materials includes mild steel. Also, tight cracks will not provide as large a break in the magnetic circuit, and will thus not exclude the field so effectively. The result is that magnetic particle techniques will be less sensitive to tight cracks, just like dye penetrant methods. The effect of the majority of crack fillings on crack detection will, however, be negligible.

4.2.3 Surface condition

Clearly the condition of the surface can have a very great effect on the detectibility of defects using visual inspection. Rough or corroded surfaces make crack-like defects harder to find without aids, and they will also affect the application of the magnetic particle or dye penetrant techniques by resisting particle motion, perhaps by masking the mouth of the defect or preventing the removal of excess liquid. Roughness may well be the result of service wear, so that the difficulties of application may increase with specimen life and hence with the likelihood that there will be defects present to be identified.

The surface of the specimen may also be covered by deposits which arise from other parts of the system. This represents a problem for local inspection of the specimen itself which has been discussed, but can also be an indirect benefit. The deposit can provide a clue to what is happening—what is being corroded or what is being worn—in other, possibly inaccessible, regions of the system being inspected. Used in this manner visual inspection can be a very valuable means of monitoring the condition of a specimen.

4.3 Ultrasonics

In any ultrasonic method there are four main components which we must consider. First, there is the way in which the specimen is stimulated, whether by piezoelectric transducer, hammer blow, heating, or by internal motion of dislocations under applied stress. Secondly, there are the nature, selection and various interactions of the elastic waves in the specimen, which determine the response to the probe. Thirdly, there is a detector, which monitors the response and which incorporates any special post-processing to extract information from received signals. Finally, there is the issue of strategy: how one arranges and selects probe and detector to do the job most effectively.

We begin with what is arguably the most popular non-destructive testing technique in current use. The basis and applications of ultrasonics are documented in the literature (e.g. Krautkramer and Krautkramer 1977). The approach relies on the propagation of elastic waves in the material under test. These will interact with defects in the material, and echoes provide an indication both of the presence of the defects and, with further analysis, their size. The key advantages of ultrasonics are the relative ease with which it penetrates materials of engineering significance, such as steel or aluminium, and the fact that it poses no significant radiation hazard requiring operational precautions. Large rotor forgings with dimensions of over one metre are regularly inspected ultrasonically. The main restriction is that a number of other materials rapidly attenuate an elastic wave (see §4.3.1). The technique is thus not so universally useful in the inspection of plastics, some heavy metals and certain composite materials, though the absence of a radiation hazard compensates for these other problems in medical applications (§3.5.1). Many large and sophisticated ultrasonic testing systems are in use or are planned but, in assessing reliability, it must be borne in mind that as much as 99% of ultrasonic testing is still carried out manually.

4.3.1 Fundamentals of ultrasonics

Resolution and attenuation: the question of wavelength
Compressional or longitudinal elastic waves will propagate in all materials whether solid, liquid or gas. Solids will also support shear stresses and the passage of shear elastic waves. Bulk compressional and shear waves are the most common modes of propagation (for general references see Kinsler and Frey (1962), Tolstoy (1973), Hudson (1980) and Graff (1975)). The velocity of elastic waves varies from material to material but the broad guides in table 4.2 may be useful. We see (figure 4.3) that to work with wavelengths in the region of one millimetre it is necessary to operate with elastic waves in the megahertz range. Since the wavelength controls the resolution of any

Table 4.2 Bulk acoustic waves.

Mode	Material	Velocity range ($m\,s^{-1}$)	Typical velocity
Compression (longitudinal or P wave)	Solid	2000–10 000	$5000\ m\,s^{-1}$ $= 5 \times 10^5\ cm\,s^{-1}$ $= 5\ mm\,\mu s^{-1}$
Shear (transverse or SV or SH waves)	Solid	1000–4000	$3000\ m\,s^{-1}$ $= 3 \times 10^5\ cm\,s^{-1}$ $= 3\ mm\,\mu s^{-1}$
Compression	Liquid	800–2000	$1400\ m\,s^{-1}$ $= 1.4 \times 10^5\ cm\,s^{-1}$ $= 1.4\ mm\,\mu s^{-1}$

technique, it would not be sensible to operate at lower frequencies for most NDT applications. In ceramics (§3.5.2) very small defects are commonly sought, so that inspection should be carried out at frequencies up to 50 MHz. For many common materials, however, the range of frequencies from 1–10 MHz is marked by a distinct increase in the attenuation of elastic waves, especially for wavelengths near to grain sizes. The result is that there may be a delicate compromise between attenuation and resolution in selecting the frequency to be used in the inspection.

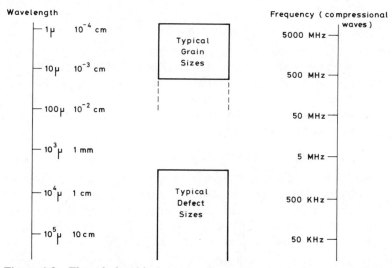

Figure 4.3   The relationship between ultrasonic wavelength and frequency and typical metallurgical features in metals.

Attenuation: general aspects

Attenuation involves three distinct components. There is scatter, e.g. by grain structure, there is absorption, as in viscoelastic systems, where ultrasound is rapidly converted into heat, and there is also beam spreading, i.e. a purely geometric aspect where intensity is lost because the detector covers only part of the beam. Normally one hopes for weak attenuation. In Al, for instance, a beam may lose intensity by a factor e in a distance of a few metres, whereas the distance may be only 0.1 m in an austenitic steel. For some purposes, strong attenuation is desirable, e.g. to suppress echoes in transducers. In such cases the attenuation may be so strong that the shape of the component may not matter, i.e. any resonances associated with precise shape can be suppressed. The geometric aspects, like beam spreading, can become complicated in anisotropic materials, including austenitic steels (Kupperman and Reimann 1978). These complications occur because it is the group velocity *not* the phase velocity which determines the speed of a pulse. Two important effects are *skewing*, when the group and phase velocity vectors are not parallel, and *amplification* when ultrasonic energy can be concentrated into a smaller solid angle than expected from simple phase-velocity arguments (see for example Tomlinson *et al* 1978 and Silk 1981). Both these features affect the precision of non-destructive testing, and hence its reliability. The examination of austenitic welds (§3.3) is an important example, since the relatively high attenuation and anisotropy of the large grains in the structure greatly complicate the analysis of the propagation of ultrasound (Dieulsaint and Royer 1980, Ogilvy 1985a,b).

Transmission through boundaries
Ultrasonic waves may be reflected from defects or from any material interface. The parameter which controls the strength of the reflection is the acoustic impedance, Z. For a given medium this is the product of the density and ultrasonic velocity. As an example, the proportion, R, of the incident energy reflected for compression waves normally incident at any interface between two media is given by

$$R = [(Z_a - Z_b)/(Z_a + Z_b)]^2.$$

The proportion of transmitted energy is thus $T = 1 - R$. The equivalent transmitted and reflected amplitudes will be proportional to $(1 - R)^{1/2}$ and $R^{1/2}$. Solids have relatively large values of acoustic impedance, with values for liquids typically smaller by about a factor of four. Gases have acoustic impedences approaching zero due to their low density, so they are virtually useless for acoustic coupling purposes.

Other factors which influence transmission are the angle of incidence, the roughness of the interface, the precise degree of slip at the interface, and whether or not contact is continuous over the beam width. We shall address some of these points in Chapter 5. Some examples of reflection, refraction and mode conversion at boundaries are given in qualitative forms in figure 4.4.

Generation of ultrasound
The commonest approach to the production of ultrasonic waves is the use of a piezoelectric transducer. An applied electric field deforms a piezo-electric plate, producing elastic waves. These can be transferred to the specimen under test via a coupling fluid, which may be a deep fluid bath or a thin film. Transfer of elastic waves between the transducer and the specimen via an air gap is, of course, totally inefficient, so the couplant is an essential part of most ultrasonic NDT. Where direct contact or the use of a couplant is not allowed, the use of non-contacting electromagnetic transducers (EMATs) or, more recently, of laser methods, may be considered, though these are comparatively rare. These various sources have been described by Silk (1984a).

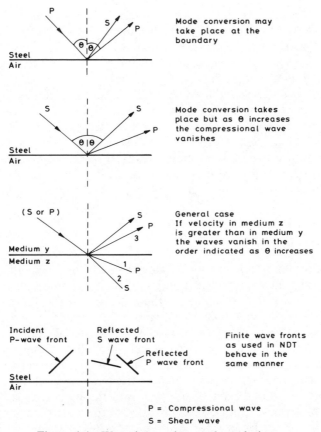

Figure 4.4   Wave interactions at boundaries.

Piezoelectric transducers may be used in continuous operation, or as a pulsed source. In continuous operation the transducer provides a piston source which gives an ultrasonic amplitude which shows diffraction effects depending on the ratio of wavelength to transducer radius. There are therefore lobes of high output, but also 'dead zones' which are not probed. Clearly defects could be missed without proper precautions. There is an obvious trade-off between high angular resolution and a survey of large regions. In pulsed operation the spread of frequencies modifies the picture slightly. The outer portion of the beam tends to come from lower frequencies, and this can produce other complications.

Electromagnetic transducers are especially convenient for operation in hostile environments or when special modes are needed and where a dipole array can be designed to generate them efficiently. They exploit magnetostriction in ferromagnets, giving mainly compression waves; the $J \times B$ force can be used for any conductor to give mainly shear waves from the force parallel to the surface. Clearly the inspection of non-magnetic non-metals, like Si_3N_4, cannot be done with electromagnetic transducers.

In thick specimens the depth of focus is important in deciding strategy (see §4.3.2). With so-called focused probes the beam is focused over only a very small depth within the specimen. A very large number of scans with separate probes will be required to cover the whole thickness of the plate. To get around this problem, Saglio and his co-workers (e.g. Cattiaux *et al* 1983) have been using very large-area focused transducers and these provide a very much larger depth of focus for measurements of this kind. The result is the achievement of high-resolution scans in thick material using C-scan imaging (§4.3.2). More recently the same kind of transducers have been used to provide time-of-flight data (§4.3.2) in addition to amplitude-based data.

The choice of mode

The most common ultrasonic waves are the bulk compressional and shear modes already mentioned. The presence of surfaces, however, allows other modes of vibration, and some of these are of use in NDT (figure 4.5). Perhaps the most common alternative mode is the Rayleigh or surface wave which propagates on the free surface of any material. Similar modes of vibration, known as Stoneley waves, may propagate along the interface between denser materials such as a solid and a liquid. Love waves are shear waves guided through a layer with particular material properties. In figure 4.5 a horizontal layer of different material properties is shown and the Love waves have a polarisation parallel to this horizontal stratum. The Rayleigh modes are localised at the surface of the specimen with amplitude falling exponentially into the bulk. They can follow surfaces through very tight bends, all round a rectangular block for instance! Thus they should not be confused with other surface-skimming waves, which are simply bulk waves travelling tangentially to the surface, and which will continue in a straight line if the surface bends away apart from diffraction effects.

Another mode of vibration is the Lamb or plate wave, which propagates in thin plate materials where the distance between the free surfaces is of the order of the compressional wave wavelength in that type of material. Whereas compressional, shear and Rayleigh waves each propagate with a definite velocity in a given material, the Lamb wave propagates at a velocity which is a function of plate thickness and of wavelength. This causes distinct differences between phase and group velocities, both of which may change as a function of the local plate thickness. This variability of velocity is a definite disadvantage in defect location unless the specimen geometry is well defined. Nevertheless, Lamb waves are used in a number of NDT examinations. Note that all waves guided in two dimensions have an amplitude falling off as $R^{-\frac{1}{2}}$, compared with R^{-1} in three dimensions from a point source. Related to these plate waves is a series of rod and tube waves which have similar general properties.

Mode conversion, i.e. the initiation of a wave of the same frequency but of a different mode, can occur whenever there is a change in density or in elastic constants. Conversion is most dramatic if the change takes place over a range comparable with or less than the ultrasonic wavelength, so an interface between two materials would be a good example. Thus, Rayleigh waves may partially mode convert to other modes at tight bends, while the surface-skimming waves may be partially mode-converted at the surface. Thus a compression wave travelling parallel and adjacent to a stress-free boundary generates a shear wave in order to satisfy the boundary conditions. This wave propagates into the bulk material at a characteristic angle and is known as a 'head-wave'. Similarly, bulk waves are normally partially mode converted at a reflecting interface, unless they impinge onto the surface.

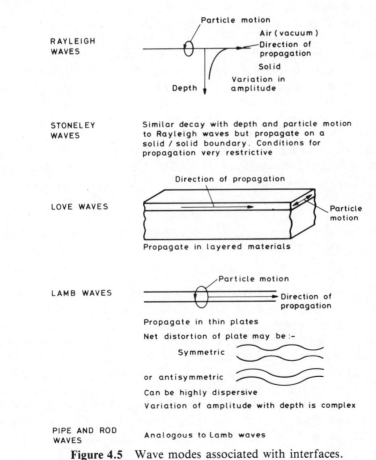

Figure 4.5 Wave modes associated with interfaces.

4.3.2 Strategies of ultrasonic examination

By the strategy we mean here the choice of ultrasonic mode, whether an amplitude or a time measurement is selected, the relative positions of any detectors and sources, and the way in which defect features are isolated. The related issues of optimum frequency, angle of incidence etc, within a given strategy, will be mentioned in Chapter 5. Silk (1982a) gives a general review.

Pulse echo techniques

The most obvious and most common ultrasonic method of defect detection is that of *pulse echo* (figure 4.6). In this a beam from a single ultrasonic probe is transmitted into the material, and the tester uses the same transducer to look for unexpected reflections which may be related to defects. To get a strong reflection, the ultrasonic beam should impinge normally on the defect surface. To this end commercial ultrasonic transducers are available providing a number of well defined beam angles in steel—commonly 35, 45, 60, 70 and 90 degrees to the specimen surface. Even so, the beam may be misaligned by 5 or 10 degrees for an arbitrary defect surface.

Generally, angled shear-wave probes are used. Here the shear wave is produced by mode conversion at the specimen surface, using an angled plastic probe shoe. This arrangement eliminates the problem of the transfer of shear wave energy across a liquid gap. With this arrangement, it is usually possible to achieve a pure shear wave, since the associated and compressional wave would be totally internally reflected at the interface (figure 4.4). Angled compression-wave probes would allow the generation of undesired shear waves, and so have been much less popular, except for time-of-flight techniques. Normal-incidence beams, on the other hand, are usually solely transmitted compression waves, hence the common misconception that shear-wave beams are always angled and compressional wave beams are always at normal incidence.

Pitch and catch

The use of a single probe in the pulse echo approach can lead to geometric problems, e.g. regions not accessible. An alternative is the use of separate transmitting and receiving probes. This makes it easier to arrange that specular reflections can be received, since the angles and positions of both the transmitter and the receiver can be separately optimised. Further, reflections may be received via the back surface of the specimen (see figure 4.6). Variants of two-probe techniques include the tandem, *Amplituden und Laufzeit Orts Kuwen* (ALOK) (amplitude and transit time locus curves) and delta techniques (figure 4.6). The ALOK approach is now becoming important in the location of vertical cracks (Barbian *et al* 1983) and is explained later in this section.

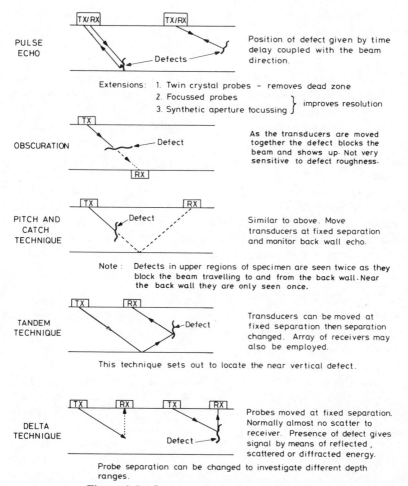

Figure 4.6 Common inspection geometries.

The advantage is gained only at the price that a relatively small volume of material is evaluated, i.e. the region common to both beams. Consequently the approach does not lend itself to a general search, with rapid scanning, unless the probable defect position and geometry is well defined and is known in advance.

As usually defined, both the pulse echo and the pitch and catch techniques rely on the reflected pulse *amplitude* to define the significance of defects detected. Defect size will be estimated from the amplitude data, either directly or through a defined decibel/amplitude criterion as the scan proceeds. Transducer/specimen coupling changes, or defect transparency, orientation and roughness variations, may combine to make the correlation between reflected pulse amplitude and defect size very weak. This is an inherent weakness with reflected amplitude approaches to quantitative defect assessment.

Obscuration

Some of this inherent variability in the techniques based on the reflection of ultrasound can be removed if the defect is used to block the beam, rather than to cause reflections. The transmitted ultrasonic amplitude is monitored, and then unexpected reductions in the signal may be used to define the presence of defects which cut the beam. The scanned distance over which beam obscuration occurs will be related to the projected defect size, regardless of factors such as surface roughness and orientation. Variations due to defect transparency and to coupling variations remain, however. The obscuration approach can be used both with a single probe and in the pitch and catch mode. Some of the problems associated with optimising the pitch and catch approach can be removed, since it is usually clear where to put the probes in relation to one another.

A-Scan, B-Scan and C-Scan

These terms describe means of presenting ultrasonic data, rather than techniques in their own right. Thus the term B-Scan could refer to data from different experiments of a wide quality range which share only this particular means of presentation.

In essence, the A-Scan (figure 4.7) is simply the trace produced on an oscilloscope after amplifying the signal detected by the receiving probe. In the normal type of ultrasonic set the pulses are usually rectified before being displayed on the screen, but any display of this kind would be referred to as an A-Scan.

The term B-Scan (figure 4.7) is applied to any presentation derived from a series of A-Scans which depicts the location of reflectors in a cross section through the specimen. Normally, the axes of the presentation would be the depth through the plate (or delay time of the echoes) and the distance scanned along the plate.

Similarly, the term C-Scan (figure 4.7) is applied to any presentation, again derived from a series of A-Scans, which shows the distribution of reflectors across an area of the specimen. Both B- and C-scans may be formed from unrectified or rectified A-Scans and both may be produced from raw data or data which may have been processed to highlight reflectors thought to be significant.

In the same way, similar terms in NDT include D-Scan, Delta-Scan and P-Scan and all describe presentations of data, scanning techniques or even items of equipment.

Time-of-flight diffraction (TOFD)

As noted, many of the problems associated with defect detection and sizing can be traced to the reliance on reflected amplitude in trying to assess the significance of defects. In an attempt to turn away from this approach, and indeed to ignore amplitude as far as possible (though clearly no amplitude, no detection!) the time-of-flight diffraction technique (TOFD) was developed within the National NDT Centre at Harwell (Silk 1977, 1984b). This relies on the time delay between the various signals to characterise defects.

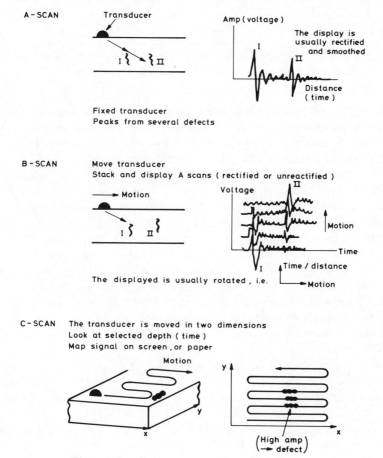

Figure 4.7   The A-, B- and C-scan techniques.

The technique can be applied in several geometries, and can use both bulk waves and Rayleigh waves. The common element is that the time delay of the echoes from the defect extremities is used to provide an estimate of defect depth. The most common geometry (figure 4.8) employs two probes astride the suspect defect position, and relies on bulk compression or bulk shear waves. These interact with the defect, and diffracted energy from the defect extremities forms coherent pulses which are detected by the receiver.

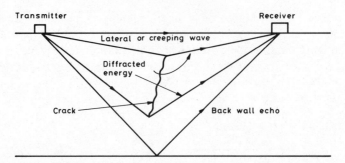

Figure 4.8 Diffraction from a crack-like defect used as the basis of the time-of-flight diffraction technique.

These pulses will be contained between others, one due to a surface skimming wave, which defines the top of the specimen, and another from a back wall echo. In this way, the time relationships between all the pulses define the size of the defect and its position within the specimen.

Bulk compressional waves are used as a rule, as the interpretation of the data is usually easier. However, shear waves appear to show up tight cracks more readily. A single-probe version of the technique has been used at Harwell (Silk 1977, 1982b) and also Gruber (1983). Automated multiprobe arrangements, using up to eight transmitters and eight receivers, have also been developed for application to inspection of the pressure vessel and nozzles of a pressurised water reactor (Curtis and Hawker 1983, Charlesworth and Hawker 1984, Stringfellow and Perring 1984). Note that the data are still interpreted interactively. Rayleigh waves can be used to provide a measure of the depth of surface-breaking defects, using the same principle of timing (e.g. Cook 1972).

Amplitude and transit time locus curves (ALOK)
This technique has some factors in common with TOFD but retains the option of involving pulse amplitude. It is essentially the application of computer control and imaging of a number of single angled-probe measurements including amplitude and time-of-flight data. The variation of echo delay time as a function of probe movement is used to select those parts of the pattern of reflected pulses which truly arrive from the defect, and these are then normally used to provide more accurate pulse amplitude data.

Holography and synthetic aperture focusing technique (SAFT)
Earlier we noted problems associated with beam spread from probes. Two alternative means of counteracting this beam spread rely on the use of ultrasonic holography or the use of the synthetic aperture focusing technique (SAFT). Both techniques can exploit computer codes which sharpen

up the recorded image from ultrasonic scans. Holography can also be carried out with analogue electronics and optical reconstruction of the data (Aldridge 1970), but this is possible only over a limited range of experimental configurations.

Holography. Ultrasonic holography is based on the same physical principles as the more widely known optical holography. In essence, a hologram is formed from the interference pattern between ultrasound reflected from the specimen under examination and a suitable reference beam (figure 4.9(a)). It can be analysed in two ways. First, it may be reconstructed to form an acoustic image, though the more practical step is to convert the acoustic hologram into a translucent grey-scale image, and to use standard optical techniques to transform this image into an optical image of the specimen and of any internal reflectors. One means of doing this has been described by Aldridge and Clare (1980). Secondly, the same kind of data collection and reconstruction can be carried out digitally in a computer, with the advantage that reconstruction of the image can then be carried out in any arbitrary coordinate system. For a long time this digital approach was very slow, and had little other advantage over the optical reconstruction route. Now, however, modern computers allow digital processing to be carried out, at a cost, in something close to real time.

Synthetic aperture focusing technique. An alternative to acoustic holography which allows the use of even more information is synthetic aperture analysis, often referred to as the synthetic aperture focusing

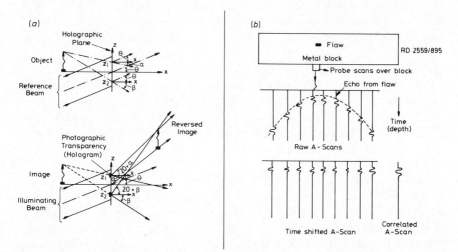

Figure 4.9 (a) Production and reconstruction of a hologram. (b) Synthetic aperture focusing technique.

technique (SAFT). The general principle is used quite widely, but it was first applied to ultrasonic data for NDT by Frederick *et al* (1978). In this approach, account is taken of the known specimen and probe geometry, and this knowledge is used to compute the manner in which the time delay of any ultrasonic echoes will vary as a function of probe movement during the ultrasonic scan (figure 4.9(*b*)). The ultrasonic data can be processed so as to reconstruct an image which reinforces data obtained near the centre of the beam and reduces data obtained near the beam edges. The result is an image which, to a first approximation, is identical to that which would have to be obtained which a much more focused probe. This reconstruction is possible even when a focused beam could not be introduced into the specimen. For general work, the effective beam width after processing is approximately one half of the aperture of the scanning transducer, but resolutions approaching the theoretical limit of one wavelength can be obtained.

4.3.3   Uses of ultrasonic techniques

We now attempt to highlight the factors involved in the ultrasonic detection of various types of defects. Clearly we cannot attempt to survey fully the enormous variety of applications and levels of success, but these are already covered in the literature. Instead we concentrate on aspects bearing on NDT reliability for the various types of defects.

In the following sections the difficulties of locating and sizing each defect type are mentioned and this will inevitably point to preferred techniques from those outlined above. However, recommendation of specific techniques is not attempted except in the broadest sense, and our emphasis is more to focus attention than to provide a reference survey.

Surface-breaking defects
Often one finds a major division in ultrasonics between the detection of those defects which lie on the accessible surface and those on the back wall of the specimen. Curiously, however, it is the latter which are most often the more easily located and sized. A wide variety of defects may exist and, as we have seen, there are a large number of specific ultrasonic techniques.

Cracks.   Crack-like defects are generally good specular reflectors of ultrasound. They may be readily detected if a near-normal incidence can be arranged. There are two basic problems, however, First, the achievement of near-normal incidence implies some control over the test geometry, and this may not exist. In inspecting for near-surface cracks, for instance, it is very difficult to obtain a good reflection with a single probe and this inspection has traditionally had a low success rate. On the other hand, a near-normal crack on the far surface can provide an almost perfect corner reflector; inspection with a single angled-probe is almost ideal for this defect. The second problem is that cracks themselves may be semitransparent, so that the amplitude of reflection may be a misleading guide to the importance of the defect. Moreover, the degree of transparency may vary as a function of the life of the structure under test, or with its loading, regardless of whether the defect has extended. In this way, repeat measurements can become confusing. It is these features and the other sources of amplitude variation, mentioned earlier, which are not related to defect size, which TOFD attempts to eliminate. The use of TOFD, however, implies a need to resolve smaller echoes than can be obtained from specular reflection. This is less of a disadvantage than it appears at first (Silk 1984b). The constraints of test geometry often mean that specular echoes are not obtained, so that even conventional ultrasonic testing will use echoes comparable with those needed for TOFD.

Wall thinning, denting.   The detection of these defects is normally only a problem if they occur on the far side of the specimen. The ultrasonic test then becomes one of monitoring changes in wall thickness, a fairly straightforward and accurate use of the technique. In effect, it is the simplest example of TOFD. Detection of such near-side defects would only be a problem if the surface was masked in some way, such as by lagging, when ultrasonics would not generally have a role to play. If the coating was thin, electromagnetic transducers (EMATs) might provide the basis of an ultrasonic test by inducing elastic waves directly into the specimen.

Surface hardness.   Surface treatments of materials can affect the local elastic properties. These effects may be monitored using ultrasonics. The changes studied could be beneficial, in which case the test would be carried out to ensure full coverage. Alternatively, the changes might be deleterious, in which case the examination is one of defect location. Many factors can affect the propagation of elastic waves, including stress, texture and composition, so that determination of some single parameter, like hardness, is not always practical. In all cases the result is an integration of the effects of the test parameters over the whole path of the ultrasonic wave, so measurements may not pick out very small areas of modified material. Usually tests are confined to sampling specimens from a well-defined production route, for instance.

Buried defects
As a rule, buried defects are of less structural importance than those which break the surface. Internal defects do not usually experience the same extremes of stress, or the ingress of water etc, leading to corrosion or to

stress corrosion. On the other hand, these buried defects can be detected less readily in general, so their contribution to failure may exceed that due to surface-breaking defects.

Weld defects. (See §3.3).   A number of defect types may be produced as a result of the welding process. Many can be regarded as crack-like, and will be treated later. Here we consider defects of a volumetric nature, such as pores and regions of slag. Such defects will be reasonable reflectors of ultrasound, regardless of angle of incidence, and may be more reliably detected than the important cracks in situations where it proves impossible to place probes optimally.

As a general rule, these volumetric defects are not so effective as crack-like defects in raising local stresses. Unless the defects are very large, they only grow to failure in highly stressed materials. Thus they are relatively easily detected and their presence can often be tolerated unless they affect the detection or growth of more significant defects.

Cracks.   When the crack is internal there are no special extra problems of detection. The internal cracks may occur at a wide variety of angles to the surface, ranging from near vertical to angles close to or near the angle of the initial weld preparation. Defect orientation (e.g. across the weld or along the weld) may also vary (see figure 3.8). This means that it is time-consuming to detect arbitrary cracks in a weld using specular reflection techniques, and difficult to automate the procedure. Recent developments, such as variable-angle probes, TOFD etc, attempt to simplify the mechanics of the scan, and may pave the way for automatic inspection.

Inclusions.   All steels and many other materials have small inclusions in their structure arising from the manufacturing process (see §3.2). Individually these are not significant defects, but, in combination, they can affect the fracture toughness of the material and its ability to withstand lamellar tearing. For this reason schemes have been suggested to monitor inclusion density in one form or another. This is not a major branch of NDT at present.

Slag Lines.   Long stringers of slag may be left behind by the welding process (§3.3) which can give the appearance of much more serious defects. These are usually volumetric defects, and are often better general reflectors than cracks. Unless steps can be taken to confirm the specular nature of the reflector, a 2 mm deep slag line may reflect as efficiently as a 10 mm crack, and, because of beam spread, will appear as deep in decibel-drop tests. This highlights the need for better quantitative characterisation of defects by NDT techniques (see Chapter 5).

Laminations. The earliest demonstrations of the potential of ultrasonic inspection showed these defects as often as cracks. Laminations are common in rolled steel plates though improvements in steel making make them rarer now. Although they may often be ignored, they will affect the strength of welds if they come to the edge of the plate, as cut to size. Laminations are ideally orientated for ultrasonic detection, and are usually the subject of automatic NDT examination at the steel mill

Manganese sulphide laminations may obscure significant crack-like defects. For this reason the ASME codes of practice for the inspection of pressure vessel components give limits on the allowable percentage of area of lamination detected by normal compression ultrasonic probes in plate material. Material with a higher area percentage of lamination is deemed uninspectable, and is rejected before being welded into a pressure vessel.

Bonding of laminates
An enormous amount of work has gone into the evaluation of ultrasonics as a technique for the determination of the quality of the bond between materials or laminates. Unfortunately, the net result has been inconclusive. Gross defects, such as the complete loss of adhesive, are easy enough to detect. Bond line thickness may be monitored too, and this is often a factor in determining bond quality. However, an effective measure of the quality of a bond line does not yet appear to have been demonstrated as a general technique, using ultrasonics or any other NDT technique. Even applications to individual bonding tasks appear to be questionable.

4.4 Impact testing

Probably the best known form of NDT is the wheel-tapping test applied to railway wheels. This is a special case of a broader range of impact testing techniques of varying complexity whose common feature is that the response to an impact is used to assess specimen properties or integrity. In some more extreme forms the effect of the test on the life of the specimen may have to be considered, and the technique should not strictly be regarded as non-destructive. There are two main areas of application, namely vibration analysis and hardness measurement. The latter is neither a method of inspection nor non-destructive (for it introduces surface damage) and so lies outside our scope.

4.4.1 Vibration analysis

A sharp impact sets the specimen into oscillation, and the nature of this oscillation is affected by the state of the specimen. Wheel-tapping is an ideal example of this type of test, since the specimens are largely identical, so that any change in vibration due to a crack can be readily detected from the 'quality' of the sound. The technique might be applied to any situation in which a large number of identical components are involved, and many schemes have been examined to automate the detection process. One stumbling block seems to be that of the reproducible mounting of irregularly shaped specimens. This seems to have limited the widespread use of the technique to the testing of wheels.

Vibration analysis can adopt non-impact forms too, e.g. using a variable frequency source, when the position of nodes and antinodes may give a clue to the presence of defects. For the purposes of this section we are concerned also with uncontrolled sources of vibration, e.g. due to ocean waves as well as impacts or sequences of impacts. With frequency analysis techniques, the use of ocean waves as sources to investigate the integrity of offshore platforms has been mooted and appears to be going through a development stage, with work aimed mainly at the collection of data.

4.5   Acoustic emission

The propagation of a fatigue crack in a specimen is accompanied by the generation of elastic waves. These may give rise to an audible response (as with tin) or, more generally, may be detected by high-frequency transducers. The idea of actual detection of crack growth as it occurs is very attractive, and has led to much work aimed at both the detection of growing cracks and the characterisation of the source of emission. Some of this work is reviewed by Stephens (1978).

In principle, the energy release can be related to the degree of crack growth, so the growth of a defect might be monitored by an integration of

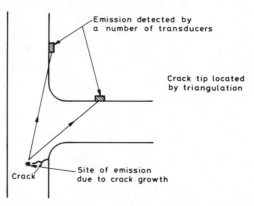

Figure 4.10   Acoustic emission for defect location.

the number of pulses and their size. This has been demonstrated on simple specimens in the laboratory (Hutton and Schwenk 1978, Scruby and Wadley 1982). However, in real technological specimens, the information in the acoustic pulse which relates to the source becomes convolved with information defining the local structure of the specimen which is often complicated. As a result, the current practical use of acoustic emission is confined to defect location (figure 4.10).

Monitoring of structures
Detection of crack growth in structures usually uses an array of ultrasonic probes. The location of the source can be pinpointed using arrival-time correlations. If a particularly 'noisy' region is found, it may be investigated in greater detail. Acoustic emission lends itself to monitoring, and data may be built up over a period of months. The main problem is the limited correlation between the growth of defects and the size of the acoustic emission. Brittle materials give rise to much greater pulses, growth for growth, and some cracks may grow in an almost noiseless fashion. Moreover, strong and consistent emission sources may be found to be due merely to the rubbing of surfaces or scale, or to coating cracking. Acoustic emission thus requires substantial back-up.

Monitoring of welding
A recent use for acoustic emission is the monitoring of welding in which one detects the elastic emissions which arise as the weld pool cools. The nature of these emissions changes if defects are present, raising the possibility of an on-line welding monitor or defect detector. The evaluation of this approach has shown that many, but not all, important defects can be located in this way at an early stage.

4.6 Thermal resistivity

Thermal resistivity can be useful in the detection of any flaw or change which prevents or enhances the flow of heat through the specimen under examination. By their nature, flaws tend to be disrupters rather than enhancers of natural heat flow, and so give rise to unnaturally hot regions if examined from the side of the heat source, or unnaturally cold regions if examined from the opposite side. However, defects such as insulation defects give rise to increased conductivity and are thus shown up by exactly opposite temperature trends.

Infrared cameras are available which can detect changes in temperature of fractions of a degree centigrade. Thermography can be used to indicate the presence of defects in any material which is conducting heat. The heat source may be naturally present, as with a boiler or chemical reactor for instance, or may be applied artificially for the test (figure 4.11). In the latter case one expects the temperature differences to be greatest during the build up of heat, rather than after steady-state conditions are established. It is then a short step to the alternative approach of pulsed thermography, in which the heat is applied as a short pulse, and the response of the specimen studied dynamically so as to isolate the time period of greatest contrast. Video-compatible infrared cameras are a valuable receiver in this application giving the pulsed video thermography technique of Reynolds and his co-workers (e.g. Reynolds and Wells 1984) (figure 4.12).

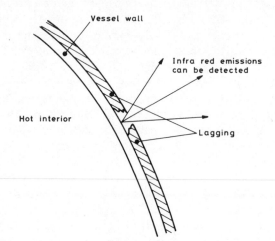

Figure 4.11   Use of static thermography to locate a flaw in thermal insulation.

Joins.   Any kind of join in a specimen is a potential site for defects, and many of these might be amenable to detection by thermal techniques. Most of these defects will be thin, e.g. lack-of-fusion defects in welds, or bond line defects (§3.3). Their effect on the flow of heat may be relatively transitory, and the heat pattern will even out given sufficient time. For this reason, pulsed heat sources are likely to be more effective than steady-state operation. This will often favour pulsed thermography, but other transitory heat sources may also be successful, e.g. inspection immediately after the welding before the weld has cooled.

Condition of concrete.   Concrete is a good example of a material which cannot be inspected by most NDT techniques. It is non-conducting, non-magnetic, and attenuates ultrasound strongly. Because it consists of an aggregate in a matrix, the conduction of heat is not very reproducible on a small scale either. However, the conductivity of heat may be expected to provide the basis of detection for larger-scale defects.

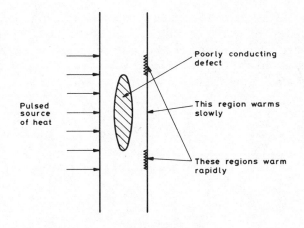

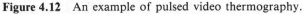

Before equilbrium is set up the region under the defect
can be clearly seen as relatively cool using an infrared video system.

Figure 4.12   An example of pulsed video thermography.

Complex materials and composites.   Composite materials, such as glass
fibre or carbon fibre reinforced plastics (GFRP or CFRP) are also very
difficult to examine by conventional means. A lot of work has been carried
out using ultrasound, without much conspicuous success. Thermal inspection techniques become very attractive for the location of many defects in
these materials, however, since the perfect composite structure involves
fibres whose thickness and conductivity are such that they do not affect heat
flow patterns greatly. The cost of glass fibre structures is much less than
those made of carbon reinforced materials, so inspection of GFRP is usually
confined to larger costly structures.

Insulation or coating inspection.   Many specimens have coverings or
coatings to insulate or protect them. The inspection of these coatings is an
important branch of NDT. Coatings which protect against corrosion are
particularly significant, since crevice corrosion may occur in the region
between the metal and the disrupted coating, and this may well be worse
than if no coating had been applied. Thermal insulation is, of course,
particularly amenable to thermal inspection, since the heat source is already
available. Similarly, many exposed pipelines carry hot or warm fluids which
may again provide the heat source for inspection. Medical infrared methods
may be regarded as a special case.

4.7   Eddy current techniques

The eddy current technique uses a coil to induce a current in the surface of
a specimen immediately below. The apparent impedance of the coil is

affected directly by the induced current. If the surface is broken by a crack or other defect, or if the conductivity or permeability of the material changes due to altered surface condition or composition, the eddy current distribution, or its density, is altered (figure 4.13). These changes can be detected by monitoring the impedance of the coil.

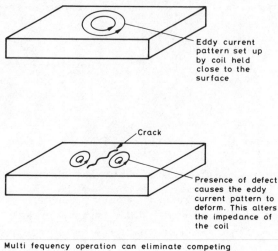

Eddy current pattern set up by coil held close to the surface

Crack

Presence of defect causes the eddy current pattern to deform. This alters the impedance of the coil

Multi fequency operation can eliminate competing effects

Figure 4.13   Eddy current inspection.

Clearly the material under test must be conducting so the technique is largely limited to metals, although work has been carried out on carbon. Further, the penetration of the eddy current is limited to the skin depth of AC electric currents in the specimen, and this limits the test to the outer layers. Waidelich (1976), Becker and Betzold (1983) and Boness (1982) give examples of eddy current methods in operation.

4.7.1   Material characterisation

It is often forgotten that NDT techniques can be used to characterise materials. One example is the use of eddy currents for metals. The conductivity of the specimen provides the clue to any change but, since many factors affect conductivity, it is not possible to use eddy currents to characterise the material absolutely. Nevertheless, one can use this approach with groups of specimens to monitor the range of a given property usefully.

Surface finish.   Since the impedance of the eddy current coil is affected by the conductivity of the materials with which it is in close proximity, a change in this level of conductivity may reflect changes in the surface finish of the substrate. So many other factors may be important that an absolute measure of surface finish would not be possible, but an eddy current technique applied to a limited range of specimens could provide an accurate indication of surface finish, perhaps indicating the desirability of further machining.

Discontinuities. A number of factors may cause an abrupt discontinuity in the conductivity near the surface of a specimen. An example would be a change from one steel type to another across a friction weld. Alternatively, a patch of material overheated by friction could give rise to an abrupt change in conductivity. In this type of problem eddy currents can be used to detect changes, even though the detection of a change does not necessarily characterise the cause of the change. Characterisation needs more information, perhaps by multifrequency eddy current analysis, to investigate the material at a number of wavelengths and depths.

Thickness of material. The depth of penetration of eddy currents is limited by the skin depth of the material. Thus induced currents in thick specimens to not 'see' the far side of the specimen and no depth estimate is possible. Clearly, however, when the material is of depth comparable to the skin depth, the penetration of eddy currents will be cut off, and the apparent conductivity of the material will change. Monitoring anomalous changes in the apparent conductivity as a function of skin depth provides a means of determining specimen thickness.

Incorrect material. One very important task in large engineering works is connected with good housekeeping: the separation of material types from one another. This is particularly true for welding rods, where each type is optimised for a rather limited number of applications and steel types. Confusion of such materials can lead to defects in the welds and, possibly, to premature failure. Eddy current monitoring of conductivity is potentially of great value in this type of non-destructive characterisation.

4.7.2   Detection of surface-breaking defects

Since eddy currents are confined close to the outer surface of the specimen under test, they are most useful in detecting defects which affect the surface directly. To these could be added those defects which do not break the surface, but which lie within one skin depth of the surface. In many metals of engineering significance, this would restrict attention to those defects within a fraction of a millimetre of the surface.

Cracks. A surface-breaking crack can block the paths of any eddy currents. The eddy currents would be forced to travel down and under the defect, if shallow, or to take up a quite new circulation pattern. In either event, the change in the coil impedance would be large and allow the crack to be detected. An estimate of the length of the defect could be obtained if the eddy current probe is scanned. The magnitude of the impedance change would be an indication of crack depth. For short cracks, where current may flow both under and around the crack, the effects of changes in depth and length will be hard to separate (figure 4.13).

Wall thinning and denting.   Both wall thinning and denting may be found by eddy current methods, but the mechanism behind their detection may not be straightforward. A narrow area of wall thinning can block the passage of eddy currents, rather like a crack. A broad area of wall thinning may not act effectively as a block, however, and may not show up, though it may be indicated through the change in the lift-off of the probe; there is then a problem of separating thinning from other causes of lift-off arising during the scan. Another mechanism for detection might be a change in the conductivity of the specimen produced by the agency which caused the wall to be thin. Examples would be work-hardening in the case of impact damage, or surface roughness in the case of corrosion. Overall, therefore, these defects may be detected well, but which actual mechanism is important needs to be defined for each case individually if estimates of the reliability of the technique are to be made.

4.7.3   Detection of buried defects

Eddy current methods have no general ability to detect buried defects. There are, however, many instances where defects close to the surface are correlated with surface changes, e.g. the effects of wear. Then the use of eddy current techniques is valid, but resolution will fall rapidly the further the defect is from the surface. A limit of detection of about twice the skin depth should be assumed.

4.8   Magnetic flux techniques

Magnetic flux techniques are based on the penetration of a magnetic field into a specimen of ferromagnetic material. If the surface is broken by a crack, or if the permeability of the material changes, the distribution or strength of the magnetic field is altered. This can be detected by monitoring the surface of the specimen by a flux meter (figure 4.14). Clearly the material under test must be ferromagnetic, so this technique is limited to relatively few metals, happily including mild steel, the most important structural material. When magnetic techniques are applied dynamically, the penetration of the magnetic field is limited to the magnetic skin depth. This limits the test to the outer layers of the specimen.

Like eddy currents, magnetic flux inspection can be used both for defect detection and for the detection of changes in the material due to surface treatment, to compositional changes or to the surface condition (Forster 1980). In defect detection, the method has much in common with the visual magnetic particle approach (§4.2.2).

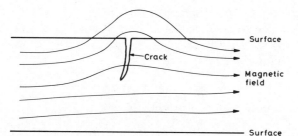

Figure 4.14 The presence of a crack forces the magnetic field out of the specimen where it can be detected.

4.8.1 Material characterisation

Here the magnetic permeability of the specimen provides material characterisation, giving an indication of any change. Once again, it is affected by many factors.

Surface finish. The condition of the surface of the specimen can affect magnetic properties. In general the effect will not be large unless it is enhanced by some extraneous effect, such as changes in permeability associated with the working of the surface. The changes in the magnetic circuit from surface irregularities will be of the order of the scale of the surface finish, and could be masked readily by changes in the lift-off of the magnet. The quality of the scan will be the major limitation on the degree of change in the surface condition which can be detected.

Discontinuities. Many of the factors which cause an abrupt discontinuity in the conductivity of a specimen surface will also affect the permeability. A change from one type or specification of material to another, a patch of material with surface cracking, a region overheated by friction, can all give rise to an abrupt change in permeability. Thus magnetic techniques can be used just like conductivity to detect changes. The same caveat applies, namely that the detection of a change does not necessarily define the cause of the change.

Thickness of material.   The depth of penetration of an alternating electro-magnetic field is limited by the skin depth of the material. When the material is of depth comparable to the skin depth, the penetration of the field will be affected. Monitoring this change provides a means of deter-mining the material thickness. This will extend to greater thicknesses than the use of eddy currents allows.

Incorrect material.   As mentioned in §4.7.1, non-destructive tests to characterise materials are an essential part of good housekeeping in a large engineering works. Clearly magnetic permeability would be one possible parameter to monitor in this connection.

4.8.2   Detection of surface-breaking defects

Static magnetic fields can penetrate deeply into magnetic materials. However, most testing situations call for a scanning approach, and the degree of movement implied by this means the field will be confined reasonably closely to the outer surface of the specimen under test. Thus magnetic techniques are also most useful for defects close to the surface, perhaps affecting the surface directly. To be detectable, internal defects must lie within one skin depth of the surface. In metals of engineering significance, these may vary from no more than a fraction of a millimetre up to several millimetres, depending on the metal properties and the scanning rate.

Cracks.   The presence of a surface-breaking crack is sufficient to block the magnetic lines of the force, which would be forced to travel outside the specimen locally. A suitable magnetic flux detector would register a local increase in field strength, and this would allow the crack to be detected. If the magnetic probe is scanned, an estimate of the length of the defect could be obtained; the magnitude of the excluded field would be an indication of crack depth.

Wall thinning and denting.   Changes in material thickness may be detected by magnetic field exclusion. Wall thinning may be found if the material is reasonably thin. It is less clear that denting would be detectable unless the dent is associated with wall thinning or some impact-induced change in the magnetic properties of the material. While dents may be detected, the actual mechanism of detection may vary from defect to defect, making uncertain any estimates of the reliability of the technique.

4.8.3   Detection of buried defects

Magnetic flux methods have very restricted ability to detect buried defects because of skin depth effects. Where defects close to the surface are associated with wear or products of surface preparation, the use of magnetic techniques is valid. Resolution will fall the further the defect from the surface, and a limit of about twice the skin depth should be assumed for detection.

4.9   Electrical resistivity

The resistivity of a specimen can be affected by the condition of the material and by its local specification. Cracks and other defects in the material provide an artificial increase in the apparent resistivity, and so may be detected electrically. In the broadest terms, this approach to NDT has the same basis as eddy current testing: the eddy current probe is simply a non-contacting means of introducing electric currents into the specimen. As with eddy current inspection, the specimen must conduct electricity. In the techniques discussed here a physical contact is made between the electrical circuit and the specimen under examination. These approaches are thus less attractive than the use of eddy currents when large areas are to be scanned, but they have advantages in controlling the entry and exit points of the current. One can constrain a current to flow under a long deep crack, for example.

Both AC and DC measurements are possible. For AC work the frequency (and thus the depth of penetration (skin depth)) of the current is at the choice of the experimenter. At first sight, it may appear that the use of DC or very low frequency AC is the most attractive means of examining defects which extend into the material. This is indeed so for buried defects. For surface-breaking defects, however, the current is forced to cross the defective region, regardless of depth of penetration. Thus a high-frequency, small skin depth current can be an accurate means of sizing the depth of surface-breaking cracks. This is in contrast to the use of eddy currents, which would simply change their pattern and would not pass under the crack in these circumstances. With electrical currents, as opposed to (say) ultrasonics, the tester has only limited control over the current paths in the specimen. Thus the current might pass around, rather than under, the defect, depending on the defect's aspect ratio. This must be take into account in the analysis of data.

Surface-breaking defects.   Surface-breaking defects may be located and sized using changes in local resistivity of a specimen. The technique will work with both DC and AC fields, but usually the latter are preferred (Dover *et al* 1979), since the resistance of the specimen is higher as the

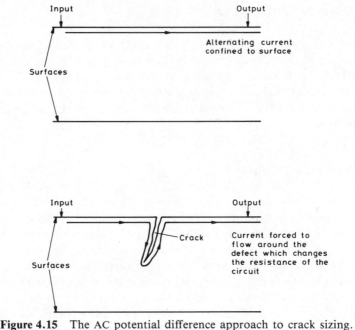

Figure 4.15   The AC potential difference approach to crack sizing.

current is confined to the skin depth (figure 4.15). Measurements are then less affected by extraneous effects and there is also less power consumption.

Internal defects.   A DC field is essential for internal defects unless the defect lies just below the surface. This requires a much more stable and more carefully set up apparatus, with the result that the technique does not appear to have been attempted outside the laboratory. In the laboratory, however, this approach appears to provide indications of the presence of defects and an estimate of their size.

Joins.   Under certain circumstances the condition of joins between conducting materials can be estimated from measurement of resistivity. If the join to be tested is a long weld, the conditions of detection are essentially the same as for defect location and sizing in general, and previous remarks apply. The condition of more restricted joins too, such as brazes and spot welds, might be determined by their ability to pass an electric current thus indicating the proportion of the join which is 'good'.

Condition of concrete reinforcement.   The condition of the reinforcement in concrete structures, particularly those in use under water, is of concern as the structures age. Reinforcement rarely breaks unless it is first weakened by corrosion effects. The state of corrosion of the reinforcement is thus of direct concern. The resistivity of the reinforcement members would seem to provide a good indication of the loss of metal, and to offer a potential solution to this inspection task. There are serious practical problems, however, such as access to the reinforcement, linkage of reinforcement, etc which mean that the measurement of conductivity is not a universal panacea.

4.10 Penetrating radiations

Any objections to the assertion that ultrasonics is the most widely used NDT technique must surely come from workers in radiography. Most certainly radiographic techniques have been in use for the longest time in recognisable NDT applications. In all cases the technique is based on using the specimen to partially block the radiation, in the expectation that the defects being sought will preferentially pass more or less radiation (figure 4.16). In this way defects will reveal themselves on film or through the use of suitable radiation detectors. Radiography may be used in inspections aimed at defect detection, intended to confirm structure, or perhaps to look for blockages in complex equipment; it may also be used to spot foreign bodies in packaged materials, including foodstuffs. The underlying phenomenon is that the object being sought modifies the local absorption of radiation, and shows up as a darker or lighter patch on the radiograph. General references to the technique are provided by Halmshaw (1966) and Chesney and Chesney (1981).

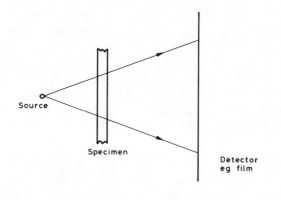

Note :– Radiography forms a shadow picture so
         magnification can be increased if the source
         size is reduced. Specialised high definition
         x–ray units allow very small defects to be
         resolved. Also the detector is usually in contact
         with the specimen.

Figure 4.16 Principle of radiography.

Four types of radiography are discussed here, though we stress that there is no essential difference between x-ray and gamma-ray techniques. Generally speaking, the more energetic gamma rays penetrate materials more readily. However, this is a generality, and there is no fixed point where a transfer in nomenclature is agreed. For practical purposes, x-rays are generally produced by an instrument and thus can be switched on and off. Gamma radiation comes from certain radioactive materials, and is 'switched on or off' only by removing or replacing shielding. Since the techniques involve some form of radiation hazard, their application is subject to stringent safety checks, especially when x-ray or gamma-ray techniques are applied on site, and procedures must be defined in advance. The operator thus has less influence on the reliability of the technique than with some other approaches. Again, the inevitable slowing effect of the safety requirements, and the element of danger, might encourage more haste in setting up detectors or films, so that some precision in the final result may be lost.

4.10.1   X-rays

X-rays cannot be conveniently focused for NDT, and so produce only a shadow picture of the specimen under examination. There is a built-in means of enlargement of the image, simply by increasing the distance from specimen to film. The limits to this enlargement and to the resolution of very small defects are provided by the effective spot size of the source. Consequently high-definition x-ray sets are being used increasingly for specialist studies (Ely 1980).

In parallel with developments in resolution there is a need to look at thicker material, which requires an increase in the energy of the radiation used. These higher energies are possible with specialised sources based on linear accelerators. Since the amount of energy which can be released in a large target will exceed that which can be released in a small target, there is a trade-off between resolution and specimen thickness: high-definition machines will be of lower energy and of lower intensity than normal sets of the same vintage. The power and intensity available for high-definition radiography is steadily increasing.

4.10.2   Gamma rays

The only essential difference between gamma radiography and x-radiography is in the manner of applying the radiation to the specimen. Gamma-ray sources are continuous emitters, and must be transported in shielded containers; exposure of the source to the specimen consists of removing part of this shield for a suitable length of time. This removal is carried out automatically. Very sophisticated sources and source holders have been designed and some allow the source to be delivered at the point of inspection and retracted with remote control.

4.10.3 Neutrons

Thermal neutrons are not rapidly absorbed by materials and can penetrate thicknesses of some inches of steel and, at present, the only neutron sources of suitable intensity are provided by reactors and accelerators, and special neutron detectors or film techniques have to be used. Studies of portable sources are in hand. Berger (1977) gives a general review.

A difference from x-radiography is noticed, however, when one looks at neutron cross sections. X-ray cross sections of materials tend to increase steadily with the atomic weight. Absorption is related to atomic weight and the density of the components. Because of this, denser or higher atomic weight materials can easily be made to stand out in radiographs, such as bones in medical radiographs or reinforcing bars in concrete. With neutrons however, there is no such steady variation in cross section. In particular, hydrogenous materials have a rather high absorption. Neutron radiography thus provides an excellent and unique means of highlighting hydrogenous structures like lubricants inside metal components. Many other elements notably the rare earth elements have high neutron cross sections, so the technique may have many specialised applications, and should be regarded as being complementary to the use of x-rays or gamma rays.

4.10.4 Protons

Proton radiography is used relatively little but has unique features which render it capable of resolving small depth changes in quite thick specimens. The idea is to use it at the limit of the proton range, where small thickness changes give rise to very large changes in the number of protons which penetrate the specimen. To make sure the end of the proton range occurs at the right spot, specimen moulds may be used to act as an initial part of the absorbing medium. The technique is not widely used, and seems unlikely to attain the level of interest of neutron radiography. For a review see West (1975).

4.10.5 Defect detection

X-ray and gamma-ray radiography produce direct pictures of potential defects, so that shape can also be used in making a decision on their nature and significance. This is not generally so in ultrasonics, even though modern data presentation techniques will now provide an image.

Recent advances in digital computer processing of radiographic data have advanced the image capability dramatically and can also be used to improve the quality of pre-existing radiographs. In these developments the computer algorithms are used to counteract the blurred effects on the radiograph which arise due to the finite size of the source. Very considerable enhancements of the image are possible using the so-called maximum entropy technique (Burch 1980, 1984), but this does require a considerable amount of processing. To some extent, therefore, financial factors will determine the degree of image resolution improvement appropriate in individual cases.

Cracks.   Here the crack must be aligned fairly closely with the beam since, seen obliquely, there is virtually no loss of mass along the beam path. For this reason radiography is not very good for detecting arbitrary cracks unless the likely location and defect type are known in advance. Satisfactory situations include radial cracks in pipe walls, where the source is at the pipe centre. There are inevitably questions about the ability of radiography to search for cracks unless very great care is to be taken. When cracks are known to exist, or are strongly suspected, then radiography may be used to clarify their nature and extent.

Volumetric weld defects.   Radiography is very good at locating regions of extra mass or missing mass. The detection of volumetric defects with a resolution in the region of 1% is fairly routine. Weld defects, such as slag and porosity (§3.3) show up in detail in most instances. If there is much slag and porosity, however, there may be problems with masking of crack records at the limit of detection. Nevertheless, this kind of defect detection problem may be regarded as being ideal for radiography.

Wall thinning and denting.   Radiography should readily locate areas of wall thinning, assuming access was available from both sides. However, a dent in a tube is not likely to involve much loss or gain of mass. The controlling factor for defect detection is whether or not significant mass has been lost from or introduced into the beam path to produce a noticeable difference in the image. Wall thinning, whether due to impact damage or corrosion, would be expected to show up fairly clearly; denting in tubes would be less likely to show except with very poor resolution through changes in the angle of the surface with respect to the beam.

Inclusions.   Most inclusions are very small, so a general radiographic technique is not an ideal way of resolving them individually. Specialised radiography, using a microfocus source, will resolve individual inclusions if required, and such methods have been used to find various very small defects in a wide variety of materials. Otherwise, radiography is more useful in determining the inclusion density as a whole, using special automatic counting techniques on the image.

Slag lines. These volumetric defects show up on weld radiographs with no difficulty. The main problem is that, by their nature, they might be confused with more worrying features like cracks, or they might mask nearby cracks close to the limits of resolution.

Porosity. Radiography using microfocus equipment has shown itself to be ideal for the task of detecting regions of microporosity. These pores are too small to be resolved individually using ultrasonics, and do not show up well even collectively.

Larger pores, with diameters from about one third of a millimetre to one millimetre or larger show up well under both ultrasonic and conventional radiographic examinations.

Finding foreign bodies in packaged goods. This is a somewhat unusual application, and is non-destructive *examination* rather than *testing*. It can be an application of major importance, particularly if the goods in question are food, or a lubricant. The detection of foreign material is possible using radiography provided firstly that the density and composition of the foreign body differs from that of its containment or substrate, and secondly that the packaging allows the radiation to enter the package.

The differences between neutron radiography and x-radiography or gamma-radiography may be important in establishing these conditions. Radiography can complement visual inspection when the foreign body cannot be detected because it is transparent, e.g. glass in a liquid of similar refractive index.

4.11 Proof testing

This is the application of extreme environmental conditions (temperature, pressure etc) in excess of those expected to occur during normal operation of the component. Since weak components fail (e.g. the failure of a cylindrical oil storage tank, 42.7 m diameter and 16.5 m high, under a full head-of-water test at Fawley in 1952, cited in Smedley (1981)) the test is not truly non-destructive and, moreover, those that survive may have reduced ability to experience the less extreme conditions of normal operation. The test is applied as a hydrostatic pressure test to pressure vessels and in a very comprehensive manner to whole F-111 fighter aircraft (Buntin 1972). The reliability of this kind of testing depends on both the test actually applied (the test itself might be omitted or misapplied due to human error, for example) and also whether the severe environment actually causes defects to grow so that subsequently they may reach a critical size during normal service (which would not have been the case if the extreme environment had not taken the component beyond its design limitations, for example).

4.12   Summary

We have mentioned an enormous variety of NDT techniques. Whilst the accent in this, and in other more detailed reviews, may seem to be on new and more quantitative techniques, it should always be remembered that the vast majority of NDT is unsophisticated, applied manually with relatively simple equipment. Such testing usually results in no permanent record other than that which the operator chooses to write. Even if the NDT techniques were themselves perfect, this procedure would rely heavily on the character of the operator and his or her state of mind from day to day. Further, most NDT relies on the operator alone to judge the importance and character of defects from an image which is itself not permanent and which comes from signals which have undergone little processing. The operator rarely has a number to read off and is often looking at pulses of some kind on an oscilloscope screen.

Even when more sophisticated images are produced by the testing system, the decision to record or to discard data is normally the operator's prerogative. The number of data which must be recorded to cover a large structure fully is so large that some reduction before storage is necessary. Yet, in a practical examination, there may be little time to make a well judged decision to save or discard. So, although attempts to quantify the reliability of specific NDT techniques are to be applauded, these data may be useless in predicting overall reliability if human factors are ignored. We shall return to this in the next chapter.

5

Failure of inspection

The iceberg theorem: only one eighth of anything can be seen

In the previous chapters we have considered the types of defect with which
we are concerned in engineering materials and the methods generally
available for detecting and sizing them. For a variety of reasons, these non-
destructive testing methods sometimes fail to detect defects of concern, or
fail to give accurate measures of defect size. This chapter examines the
reasons why techniques fail and the ways in which their limitations can be
minimised.

We must first specify what we mean by reliability of non-destructive
testing and then we may discuss how 'unreliability' occurs in practice. We
shall follow Haines *et al* (1982) who have discussed the concept of reliability
of non-destructive testing. They defined a *reliable* non-destructive testing
technique as 'one that, when rigorously applied by a number of teams, consistently detects all defects of concern'. The basic ingredients of 'reliability'
are all included in this definition.

Although most of the examples discussed in this chapter are taken from
ultrasonic or radiographic examination of steel, many of the important
points and concepts are valid more generally, and carry over to other non-
destructive inspections of different materials. Apart from the physical
characteristics of the defects of concern, the reasons for failure of inspections are largely the same in any system. Human errors or badly written
procedures can affect any technique and are, on the whole, independent of
the material being tested. The same is true also of the relationship between
manual and automated scans and the value of repeat inspections. The comments which we make on materials or the physical characteristics of defects
are, of course, specific to the ultrasonic inspection of steel, but the ways in
which errors occur and how they can be best controlled apply much more
generally.

5.1 Reasons for failure of NDT techniques

Any non-destructive testing technique consists of several distinct elements.

These and their mutual interactions are illustrated in figure 5.1. Here we single out the following components.

(*a*) An inspection technique and strategy of operation such as any of the approaches discussed in the previous chapter.

(*b*) The physical embodiment of this technique as equipment—the hardware or kit.

(*c*) An inspection procedure, i.e. the set of rules under which the kit is used to scrutinise the component under test.

(*d*) A means of application of the kit to the component. In the case of a manual inspection, it is applied to the component by a human being, or, in the case of an automated inspection, by a machine.

(*e*) The keeping of records. The raw data may be recorded, although this step may not be present for all of the data in either manual or automated inspections, and data will often be heavily edited.

(*f*) The interpretation of recorded results in terms of defect indications and characteristics.

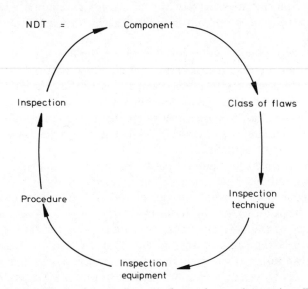

Figure 5.1 The basic components of non-destructive testing. Each of the aspects shown corresponds to a decision which must be made in devising the best approach.

Anyone faced with the task of deciding whether or not to believe the results of a non-destructive test should ask at least the following questions.

(*a*) Is the technique appropriate for the defects sought?

(*b*) Has it been previously demonstrated?

(*c*) Is the procedure satisfactory and has it been correctly applied?

(d) Did the equipment work properly?

(e) Has enough thought been given to the ways in which human error could have influenced the results?

(f) Are there any properties of the defect which have not been considered and which might impair the performance of the technique?

We shall work our way through this list, adding a little more flesh to the bare bones of the outline above.

5.2 Is the technique appropriate for the defects sought?

In order to answer this question, we must understand the physical principles of the technique itself, and be certain about the properties of the defect which we are seeking.

Clearly the technique must be appropriate in some way. When we choose among the common NDT techniques discussed in Chapter 4, we must make sure that the physical characteristics of the defect sought can influence the physical quantity measured by the technique.

The defects are, however, not under the control of the operator (cf Chapter 3). In reviewing the cracks and leaks which had occurred in the previous 15 years in boiling water reactor and pressurised water reactor piping over 4 inches in diameter, Bush (1979) noted that since the defects (stress-corrosion cracking, thermal fatigue, construction defects and erosion–corrosion) had all propagated from the inner surface of the pipes they would not be detected with a consistently high reliability. He concluded that conventional ultrasonics relying on amplitude alone would not lead to a consistently high reliability of flaw detection and suggested that the frequency spectrum be investigated for more information. Here the defect nature made it necessary to go beyond the standard methods.

5.2.1 Knowledge of defects sought

We have already described the common defects of concern in non-destructive testing in Chapter 3. Here we note that knowing about the defects one is looking for, although necessary, is not sufficient. For example, on a combat aeroplane of the Royal Air Force (Kent 1977) it was known that the critical defect was a crack of size 0.8 mm. The crack was likely to occur under the head of a countersunk fastener. Even knowing what to look for and knowing where to look is only the first step in solving problems of this sort and, in the event, an aircraft was lost due to wing failure. A typical problem in the inspection of airframes is to detect and correctly size cracks down to very small sizes, often less than 1 mm, in any of a large number of fastener holes, say up to 2000 per aeroplane. Kent concluded that such demands were 'quite beyond present day capability'. As a second example Haines (1977) pointed to the inadequacy of the American or German national codes for the ultrasonic inspection of nuclear reactors for the techniques and procedures in effect at the time. In particular, to detect a crack running directly into the wall normal to the surface, using a 60 degree shear-wave probe, a threshold of at least 20 dB below that of either code would be required. As a result the tandem technique is now a vital part of the reliable detection of certain planar defects, since this technique is optimised for these particular planar defects orientated perpendicular to the pressure-retaining surface.

5.2.2   Limits of technique capability

There are limits imposed on any technique, even though it may be quite suitable in general for the inspection required. For example, Serabian (1982), amongst others, argues that orientation is an important defect characteristic governing detectability for subsurface defects. For defect misorientations of greater than $8°$ there is a maximum defect size which amplitude-based ultrasonic techniques will detect at the sensitivities laid down. This is in agreement with the work of Haines (1977).

The accuracy of determination of flaw size by ultrasonics was discussed in general terms by Erhard *et al* (1979), who compared the resolutions and accuracies of holography, of focused probes and of deconvolution methods, and who noted that the size of the defect can only be determined to some sizeable fraction of a wavelength. The resolution determines the smallest size of an individual flaw which can be measured adequately.

5.2.3   Optimisation of NDT technique

Once we have chosen a technique which measures a physical property closely correlated with the aim of the technique (for example, ultrasonic waves might be used to detect crack-like defects since the stress discontinuities caused by defects scatter the waves strongly) we should then optimise it. This should provide us with a technique which has the desired *'capability'* to detect the defects of concern. This capability will be affected by our success in choosing optimal frequencies, angles etc, as well as by what we know of the defects (see Chapter 3), by how well we select a method (Chapter 4), and by the potential importance of the defects. In §5.10, we consider the properties of the defects which may influence the confidence we can have in the results of the inspection. We note that the defects sought should be specified with reference to some sound physical basis. This problem was discussed by Bush (1981) who suggested that the limits for flaw detection should be determined by some sensible criterion related to the severity of the flaw and the likelihood of failure of the component. Such criteria can be found, for example, in fracture mechanics.

5.3. Inspection procedures

Suppose a technique with the necessary capability is applied to the component under test according to a set of rules. These rules constitute the *inspection procedure*. Many important standard codes of practice exist. A list of NDT standards and procedures current in America in 1980 for eddy current, radiographic, ultrasonic, magnetic particle, and dye penetrant methods is published (ASNT Publications Committee 1980) and is revised from time to time. Of particular interest to us are the various British Standards (1983). These and other national standards, for example the ASME code for the construction and inspection of pressure vessels (ASME 1974, 1977, 1983—the most recent addition supercedes the earlier versions) lay down rules on the type of transducer, the angle to the normal to the surface of the component that the ultrasonic beam makes, the centre frequency and bandwidth, tolerances and so on. A survey of the codes of practice and procedures used by the aerospace industry in Australia is given by Bott (1983). These procedures or 'codes', also specify calibration methods and usually lay down a scanning raster on the surface of the component.

There are many ways in which a procedure can lead to unreliable inspection. Amongst these, we note that equipment might not be adequately specified, or some of the tolerances allowed may be too large. The coverage of the defective area might be poor due to a scanning raster which is too coarse. Some variable such as temperature may have a dramatic effect, but could be omitted from the specification of the calibration required.

5.3.1 *Effects of material properties*

Procedures may need to distinguish between different physical properties of the material making up the component to be inspected. For example O'Brien *et al* (1975) and Packman (1973) considered the critical defect size in a number of different aluminium alloys used in the manufacture of aeroplanes. For a particular geometry and applied load they found critical crack sizes of 4.0 mm in 7075-T652 alloy compared with 25.0 mm for 2024-T651 alloy. Despite this obvious difference in the size of cracks which *need* to be found in the two different materials, the current inspection procedures do not distinguish between them.

5.3.2 Changes to procedures

Changes to procedures should always be considered if the existing ones are shown to be inadequate. There are dangers though, and the new procedures must be put through a rigorous series of tests before being allowed to replace the existing (and presumably well tested) procedures.

An example of some of the pitfalls which can occur when introducing a new ultrasonic inspection code is given by Taylor and Selby (1981). They compared the ASME code requirements of the 1977 edition with the 1974 sensitivity levels. These codes make use of a *distance amplitude correction* or DAC in which the ultrasonic signals from three side-drilled holes at different depths in the calibration block are used to set a reference level of signal. Any signal displayed on the oscilloscope screen is compared with some preset fraction of the amplitude against distance curve or DAC. Different percentages of DAC are used to determine whether defect indications are merely reported or must be reported and evaluated. The sensitivity requirements of the 1977 edition were not sufficient to detect minimum flaws as laid out by the same code. Taylor and Selby found that 26% of flaws evaluated produced a response greater than 100% DAC and would, therefore, have been evaluated. However, 64% of the flaws produced a response greater than 50% DAC so that, based on the 1977 edition, these would only have been recorded and not necessarily evaluated. The 1977 edition allows a notch as a reference reflector but this reflector produces a larger signal than the side-drilled hole of the 1974 code. Therefore, the sensitivity of the 1977 code is about 6 dB below that compared with the 1974 code for a wall thickness of 0.4 in and 16 dB below for a wall thickness of 2.4 in. The standard deviation found in these tests was about 2 dB.

5.3.3 Role of modelling in the design and validation of procedures

Modelling can be used effectively to check whether inspection procedures have been designed to provide the maximum inspection capability available from a given technique.

Consider, for example, the likelihood of missing a defect with x-rays (§4.10), given the defect parameters and the geometry of the component. As a specific example, think of a penny-shaped crack of radius a and thickness t, buried in a parallel-sided plate of thickness D where $D \gg t$ The normal to the defect lies at an angle θ to the normal to the plate surfaces. A detector of radius r is positioned centrally below the defect (since this is the most sensitive position of the detector it is the only one we need to consider) and a source of x-rays is positioned above the defect (see figure 5.2). The response Φ of the detector relies on the difference in absorption between flaw and host. For $r > a$, Φ can be written as:

$$\Phi = \eta N \pi (r^2 - p^2) \exp(-\mu D) + p^2 \exp[-\mu(D-q)]. \qquad (5.1)$$

Here η is the intrinsic efficiency of the detector, N is the photon flux (per $cm^2 s^{-1}$) and p and q are geometric parameters specified by

$$p = a \cos \theta + t \sin \theta \qquad (5.2)$$

$$q = -2a \sin \theta - t \cos \theta. \qquad (5.3)$$

For $r \leqslant a$ we have a simpler expression for Φ:

$$\Phi = \eta N \pi r^2 \exp[-\mu(D - q)] \qquad (5.4)$$

where μ is the coefficient of absorption of the parent body (metal, cement, polymer or whatever). With this relationship before us we can tell what photon flux is required for a given material and likely defect dimensions and orientation in order to yield a specified signal level. Alternatively we could show what orientations would be acceptable, with the photon flux normal to the component surface, in order to provide detectable defect indications in the material, again for known defect parameters. Similarly, for a range of defect orientations we could use this modelling approach to work out how many orientations of source and detector would be required to give an adequate detection capability for these several defect orientations. These variations would be written into a radiographic inspection procedure. Alternatively, if we are given a radiographic inspection procedure for which the modelling procedure outlined above is relevant, we could predict which defects would be difficult to detect ($\theta = 0°$) and those which would be easiest to detect ($\theta = 90°$) and by how much these signals differed in the two cases.

The ultrasonic inspection procedures laid down for the inspection of a British pressurised water reactor have also been analysed by modelling (Coffey *et al* 1982). Here it has been possible to identify and correct a number of weaknesses in the inspection procedure from a knowledge of the

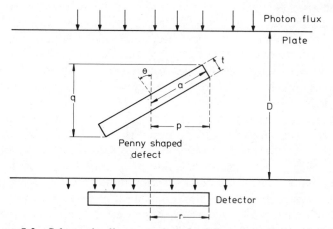

Figure 5.2 Schematic diagram of x-ray defect detection and sizing. Note that in practice the radiation is usually diverging.

diffraction of ultrasound by the edges of a postulated planar, crack-like, elliptical defect at a number of possible positions with a number of transducers scanning over the surface regions of the pressure vessel. Whilst our radiographic example was almost trivial, it is far from trivial to justify the correctness and applicability of inspection procedures employing many different ultrasonic transducers in regions of some geometrical complexity without recourse to both experimental trials and theoretical modelling. This combination of theory and experiment is likely to be an area of increasing importance in all areas of non-destructive testing as the complexity of procedures and the sophistication of inspection techniques increase.

5.4   Automated versus manual scanning

There is a difference in emphasis between an automated inspection and a manual one. In an automated inspection, we must be confident that the machine has been designed to perform the task properly and that it has been correctly set up on the component. This means that the design should have correctly accounted for the geometry of the component, and, for example, probes do not lift off the surface as scanning proceeds. Nor should there be excessive wear on the probes, or inappropriate materials be used (e.g. corrosive fluids should be avoided). Whether or not the kit has been set up correctly on the component depends on manual skills, i.e. on the ability of people to perform tasks without making mistakes. This topic is discussed more fully in §5.5.

Manual inspections are still very common. With the advent of robots, the number of manual inspections may decrease, but it will decline only slowly. In a pressurised water reactor, for example, about 90% of the inspections must be carried out manually. The reason for this is that many inspections must be carried out on small, odd-shaped, or inaccessible parts. Programmable robots may be used to carry out some of these tasks but often it is not possible to have either enough robots programmed differently, or time to teach robots the various different scans which are necessary on a wide range of components. In the aerospace industry there is also a considerable commitment to manual inspection.

Improvements often result from automation of manual scanning. An example of this for ultrasonic inspection was reported by Lewis *et al* (1979). They concluded that, at that time, the American Airforce were unable to achieve a 90% probability of detection of a crack half an inch long with 95% confidence limits using manual techniques. Changing from hand-held transducer scans to semi-automated inspection performed with a transducer-positioning device markedly improved detection reliability. This suggested that the specified reliability criterion could be achieved for crack sizes considerably smaller than half an inch long. They also reported that significant variability between operators was observed, but had not been quantified.

5.4.1   Improvements possible with automation

Some feeling for the effort involved in aerospace inspection, for the improvements which have been made and for future prospects can be found in the article by Zambon (1971). Although rather old, the article does put forward ideas for future improvements in aerospace NDT which have yet to be fully developed. In particular, one idea is that of airborne monitoring equipment for such things as engine oil, engine and airframe condition which can be telemetered back to the operating company's maintenance depots. One system along these lines, the monopulse secondary surveillance radar (SSR) at present being installed at airfields in Britain, will be replaced in the late 1980s by a system currently called 'Mode S'. With Mode S linked into the flight management computer, the ground station will be able to obtain far more additional information than is currently available. A running commentary on the mechanical state of the aircraft's airframe, equipment, engines and aviation electronics systems may be extracted during the flight, so that maintenance services and spares could be marshalled before landing. Additional information about the aircraft's weight, number of passengers, and the volume of freight it is carrying could, in an ideal world, be used to improve airport management and the use of terminal buildings (Reed 1985).

As an example of the difficulty of designing automated inspection equipment for each task which arises in the course of in-service inspection of aero-engines, Zambon (1971) notes the example of the development of root cracks in the sixth-stage compressor blades on the Pratt and Whitney JT8 engine used on the Boeing 727. Inspection for these defects meant either taking the entire engine apart at cost of 155 person hours and 30 hours elapsed time, or developing a method for inspection *in situ*. This is the sort of task for which free-standing, programmable robots would be ideal. It is impractical and uneconomic to design specific automated inspection techniques for all of these inspections and as it will be some time before robots are programmed for all the different tasks, manual inspections will persist for some time to come.

5.5   Human error

Human beings make mistakes: 'To err is human, to forgive, divine' (Alexander Pope in his *Essay on Criticism*). Whilst mistakes are only detrimental to the reliability of an inspection if they go undetected, we must take an interest in the likelihood of these mistakes and take precautions to reduce their potential for harm.

Suppose that the inspection technique is capable (§5.2.3) of the task in hand, and that the procedure is sound and has been extensively tested. There have been many successful applications of the technique. Yet something goes drastically wrong. The managing director is searching for the reason. After several gruelling interviews by successive links in the management chain one of the possible answers could be that Basil Bodger was to blame. The particular item under scrutiny may have passed through the works early on a Monday morning or late on a Friday afternoon, possibly paying the price of craftsmanship with a hangover or in a hurried attempt to get off on holiday. Perhaps the trained inspector was off sick, and Basil Bodger stepped in to keep production moving. Possibly access to this particular item was very difficult, or it involved working under water, or in a very hot room, or even in a radioactive environment. Management may have been ultimately responsible because, perhaps, the task was scheduled on too tight a time-scale, or the person overseeing the work put the inspector under too much stress. These examples illustrate but a few of the opportunities for human error.

Another possibility is that the inspector was curious about the equipment and twiddled some of the knobs. Even in the case of automated equipment, it is usual for there to be human intervention at some stage, as when the equipment is set up. In one round-robin trial, an automated ultrasonic scanner requiring water immersion for coupling was set going under computer control for an unattended overnight scan—but with no water in the tank!

5.5.1   Interpretation of results

Results have to be interpreted, and this may involve differing amounts of human intervention or expertise. An automated method of interpretation may be used, but even this will probably have been developed through some tests in which machine interpretation was compared with that of a human being. The machine is trained according to some algorithm, and pursues this scheme more diligently than humans would. It therefore tends to give more consistent results. However, the algorithm could be faulty giving the *wrong* results consistently. Likewise, with purely human interpretation we expect to have problems of human error, variability and subjectivity.

5.5.2   Human intervention

Some of the causes of unreliable inspection created by human intervention are: incentive; integrity; inquisitiveness; ingenuity; and procedures.

Considering the incentive, it is bad practice to employ a person to find defects and then to create barriers for them in reporting those defects found. In other words, if the idea is to find defects, don't make it difficult for the inspector to report them. The bright light in an inspectors eyes and a gruelling interview with a superior along the lines of 'How can you be one hundred per cent certain that this defect exists? How big is it *really*? You know that if we dig in there and we don't find anything then you can look for another employer' is hardly likely to lead to the inspector reporting many defects. How many he or she finds and fails to report will be a different story. Thus we must consider not only the inspection but also the management of the inspection.

Note, however, that human intervention is not always bad. A well trained expert can influence an inspection to the better by extracting small signals from a noisy background, or by making intelligent judgements on the characteristics of defects, for example. These good effects are just as hard to quantify as the bad ones. We do not need to worry too much about improvement in the inspection technique because of human intervention. By definition, improvement does not make the technique unreliable. It is those actions which make an otherwise reliable technique unreliable which we are seeking to identify and eliminate as far as possible.

An example of the effect of incentive is documented by Lewis *et al* (1979). In a round-robin exercise involving eddy current, dye penetrant, radiographic and ultrasonic inspections, one American Airforce base out of the 21 bases involved in the crack detection exercise, achieved a standard much higher than the others. The same equipment was used and all the technicians were of the same overall level of skill, since they were all trained and certificated in the same way. There were no differences of formal education or NDT experience. However, these particular operators exhibited a high degree of motivation: they were regarded as specialists, and had been carefully selected from volunteers.

The integrity of inspectors is essential, for otherwise there will be no guarantee that unacceptable defects will be reported, even if detected. Whilst the inspector is expected to have high integrity and demonstrate a high level of skill, there should also be some safeguards against over-inquisitiveness. If equipment is designed with controls which should not be reset during an inspection, then these controls should be hidden behind a removable cover. If they are left with too easy access, an inquisitive inspector (or a wandering meddler) may be tempted to twiddle them with the possibility of upsetting performance of the equipment, or indeed rendering it outside the standards laid down. As well as inquisitiveness connected with unnecessary control knobs, there is the possibility that an inspector may devise *ingenious* modifications to the test procedure. If these modifications are successful, and the technique improved thereby, then credit is due to the inspector. A more likely outcome will be that the modifications remain unverified at best, and at worst will impair performance. All modifications to test procedures must be thoroughly tested and documented and the inspectors fully trained in their use.

This applies, of course, to the inspection procedures when they are first developed. A common failing of procedures is that they do not convey to the inspector what was actually in the mind of the person who devised them. The only sure way to test them is to try them out on a large sample of people and observe how successful they are. A suitable criterion can be devised in terms of, say, the level of success of interpretation achieved during the first or second encounter. Thus, in terms of ensuring that inspectors carry out the right tasks, we might say a procedure is acceptable if out of a population of inspectors 95% achieve success on the second encounter. The second encounter is usually a better measure than the first since we assume that the inspectors will undergo extensive training which will allow several encounters with the inspection procedures.

5.5.3   Types of error

The most important parameters are those with the largest influence on the reliability of the inspection technique, i.e. those with the greatest effect on the spread of the results. As we shall see later in §5.6, this spread is defined by the standard deviation of the results. In talking of standard deviations, we imagine the sort of distribution in results which follows some smooth distribution, such as occurs in the normal carrying out of any experimental measurement. However, there is another sort of error which may not be distributed according to nice smooth distributions. In this category could be really serious errors which have nothing to do with the inspection technique itself. An example would be if the inspector, in preparing a final report, lost the record of a serious defect and so did not include it. This is not an error which should occur very often but its consequence could be very serious.

It is often convenient to split errors up into at least two classes, namely *errors of kind* and *errors of degree*. The terms 'errors of kind and degree' were coined by Coffey (1982) following others (Bogie and Beyers 1982, 1984) who described errors of kind as 'level 1' and 'level 2' errors as errors of degree.

Errors of kind are those rare errors due to events beyond the control of the non-destructive test itself (like loss of data) which generally have large, or serious, effects. Such errors can be best controlled or reduced to very low levels of incidence by proper procedures and suitable training and supervision of inspectors. Such errors can be estimated to some extent, and experience gained in test-block trials should result in improved procedures (see §5.8).

Errors of degree generally have a small effect in isolation but may be compounded many times to impair the performance of the non-destructive test. These errors (see §5.6) are those usually associated with experimental observations, such as taking readings with instruments, and hence can be measured with repetitive experiments on each important variable with others kept constant. Some may be the result of a combination of 'simple' errors. Each of these simple errors might be estimated easily, e.g. reading a voltage from an oscilloscope screen, and it may be simpler to arrive at a value for the spread of the overall error of degree by performing measurements on each simple error.

5.6   Magnitudes of errors in manual inspection

This section discusses errors of degree, as defined in the previous section. In §5.8, estimates of the more serious errors of kind for manual or automated inspection are discussed.

5.6.1   Distribution of practical results

The question of whether to attribute a failure of the inspection to the procedure used, or its actual application, or to a human error is not always easy or indeed useful. Some of the material in this section could equally well be discussed in §5.8. However, there is a distinction between the human error discussed later which refers to gross errors, and those which would be regarded as experimental measurement errors.

Each inspection can be regarded as a measurement which, if many such measurements are made, will have a distribution about a mean value. Even in radiographic inspections there will be variability associated with setting up the radiation source and film, the film itself (presumably small and well documented variability), and the interpretation of the developed film. In the interpretation of the developed film, the spread we are going to consider would arise if several inspectors were to separately consider the same image. For other inspection techniques the spread in this distribution is a measure of the accuracy to which the procedures can be followed in practice, that is a measure of their *repeatability*. A diagram showing the likely spread in results due to various causes is given in figure 5.3 and some typical results are listed below. In the figure, the various published results on mean and standard deviations of errors introduced by different aspects of manual inspections have been interpreted as though the errors were normally distributed (see the Appendix). This is a reasonable assumption, although there are usually insufficient data to distinguish between different types of distribution. The figure is then a composite plot of all the different errors

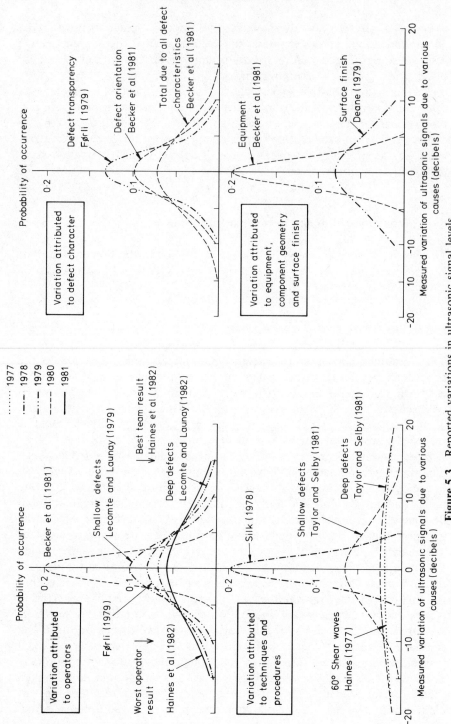

Figure 5.3 Reported variations in ultrasonic signal levels.

and their spreads, and it aims to draw attention to the typical magnitude of the mean error and the likely spread observed in practice.

From a round-robin exercise in ultrasonic inspection involving about 1500 test sample measurements Doctor *et al* (1982) concluded that there are large differences in performance between different inspection teams all using the same nominal ultrasonic procedures. For clad ferritic pipes, an indicator of team effectiveness was found to be the care and accuracy of plotting of the axial position of indications. Doctor *et al* also found that there seemed to be little difference between ideal laboratory conditions and those where access was more difficult. This last conclusion seems rather difficult to accept. It disagrees with observations on the PISC I results (Haines *et al* 1982), (PISC originally stood for Plate Inspection Steering Committee but now stands for Programme for Inspection of Steel Components). However, Doctor *et al* (1983) have restated the conclusion, together with more detailed analysis of their results. Another measurement of this variability in manual ultrasonic inspections was reported by Becker *et al* (1981) who found a standard deviation between inspection teams of 2 dB.

In the case of eddy current inspection of bolt holes to discover fatigue cracks in 7075-T6 aluminium alloy, Knoor (1974) found a variability in the detection probability of defects between nine operators of as much as 0.2. This means that some operators were achieving probabilities of crack detection as high as 0.95 compared with others who only achieved 0.75. This figure is for small cracks of area less than 0.3 mm^2. For cracks of area greater than 3 mm^2, all but one of the operators were achieving a detection probability better than 0.95.

5.6.2 *Improvements in manual inspections*

The variability in results from manual inspections can often be reduced by using more than one inspector and repeating the inspections. A measure of the variability of the results is given by the standard deviation of the measurements. Førli (1979) reported standard deviations in ultrasonic echo amplitude distributions of 5 dB due to the operator and 3 dB due to the characteristics of the specimen or defect. This figure of 3 dB variation in signals from defects refers to defects of the same type. Defect types considered in these tests were lack of root penetration and lack of side-wall fusion. Førli's results give a total standard deviation of about 6 dB equivalent to an uncertainty in defect height of a factor of four at 95% confidence limits when determined by a single ultrasonic operator. This uncertainty decreases by a factor of two when large numbers of independent inspectors are employed. Førli also found that there was a difference of 10 dB between the 'best' and 'worst' operator.

Hansen (1983), presented his view on one way of improving the reliability of manual ultrasonic inspection. This is to make the operators plot the number of defect indications against the defect size on probability paper. Each welder has his or her own 'fingerprint' in terms of the number of defects expected in some length of weld. Plotting the inspection results on probability paper allows a comparison to be made with the expected defect rate of the welder and, in addition, this technique helps to overcome some of the rare events or outliers (see the Appendix) which are the cause of many failures in detection and sizing. Because these odd events stand out when plotted on the probability paper, there is much more chance that the measurement will be repeated and bad welds identified satisfactorily.

5.6.3   Effects of procedures and team performance

Good examples of the results of inadequate ultrasonic procedures were identified by the PISC I results. These results are reported by the Plate Inspection Steering Committee (1979) and are discussed in Chapter 6. Whittle and Coffey (1981) discussed the PISC I exercise itself and its relevance to the ultrasonic inspection of pressure vessels of a pressurised water reactor. They argue that 'the PISC I ultrasonic procedure was a poor one whose failure was predictable'.

In discussing these results, Haines *et al* (1982) plotted histograms of peak signal amplitudes recorded amongst teams. These show that there were a certain number of teams who failed to report defects at the 50% DAC level, and that amongst those who did report the defects there is a large spread in peak signal amplitude recorded. This range is from 50% DAC to over 300% DAC. The mean of all teams was about 95% DAC. Some of this variability in team performance can be attributed to the amount of effort spent on obtaining a good calibration signal compared with the effort spent on obtaining signals from defects. If a large effort is put into the calibration signal it will tend to give a large signal and it will be difficult to obtain signals as large from the defects and vice versa. The conclusion can be drawn that teams have a characteristic rating, a good team tending to obtain large defect signals from all defects whereas poor teams consistently obtain small signals.

The opposite point of view was taken by Jamison and Dau (1981) and Jamison and McDearman (1981) who found that there appeared to be *no* difference between probability of detection between teams. They stated that, under the test conditions considered, there was no significant operator effect. That is, they argue that essentially the same results can be expected from one team inspecting the block several times and several teams inspecting the block one or more times each.

5.6.4   The importance of careful design of inspection procedures

Many factors can influence the performance of an inspection technique, and it is only too easy to overlook some of these inadvertently. An example from ultrasonic inspection suffices. Golovkin (1973) reported on the effect of scanning speed and spacing on reliability of results of ultrasonic inspection of tubes. The point made was basically simple: the scanning speed must be determined by the pulse repetition rate and by the minimum size of defect which is to be recorded in a given volume of metal. If the scanning speed is too fast compared with the pulse repetition rate, then the input pulses will be delivered too sparsely and large amounts of the material will go uninspected. Although this seems obvious, we can assume that the article was published because a mistake of this sort seemed inevitable: choosing an inappropriate scanning speed was foreseen, or had perhaps occurred.

5.6.5   Other factors influencing manual ultrasonic inspection

Another important variable in inspections, particularly with ultrasonics, is the coupling medium used (§3.3). Often variations in the surface topography will result in variations in the coupling of the inspection technique to the sample under investigation. For ultrasonic inspections a typical value of the unreliability due to coupling variations is a standard deviation of 2 dB (Silk 1978).

5.6.6   Combinations of errors

In the previous sections and in figure 5.3 we have presented information on mean errors and associated standard deviations from the mean. In fact the mean errors in signal amplitudes are zero since there is usually no absolute reference. The signals have been assumed to be measured on a logarithmic scale, that is in decibels, and the errors are assumed to be normally distributed on that scale. To interpret these results, we assume at first that a measurement is made which involves only one possible source of error. A signal level S is measured. In a large number of repeated measurements the signal level would lie in the range $S \pm \sigma$ in about 68% of cases and in the range $S \pm 2\sigma$ in about 95% of cases. If more than one source of error is possible, each with its own standard deviation σ_i, and we make many measurements of signal amplitude, then there are a number of different possible outcomes depending on the way in which the errors combine. In many cases the errors can be treated as independent, then about the mean signal level there will be a distribution of values with a combined standard deviation σ given by

$$\sigma^2 = \sigma_1^2 + \sigma_2^2 + \ldots + \sigma_n^2$$

for n sources of error. Cases where the errors may not be treated as independent are much more complicated, and reference should be made to standard texts such as Kendall and Stuart (1977, 1979, 1982).

5.7 Equipment malfunction

Equipment does not work relentlessly for all time, but wears out and needs replacing. These malfunctions will lead to an unreliable non-destructive technique, though it will only be a problem if it goes undetected. The solution to this problem is redundancy in the equipment and cross-checking including carefully laid out procedures for use of equipment (see §5.3). Equipment should be checked to see that it is working correctly at the beginning and end of a test as a minimum requirement. The procedure should lay out the tests which are required in order to test whether malfunction has been a problem. Logs of equipment breakdown should be kept so that it is possible to estimate what level of effect they could have if they were not noticed at the time.

Another problem introduced by equipment breakdown is that of repair or replacement. The part of the kit replaced may not have the correct specification, or it may affect other parts adversely so that the kit as a whole no longer meets the required specification. Clearly, if parts are replaced in the inspection kit during a test, then the whole setting-up procedure should be carried out again.

We have mentioned that redundancy in equipment is part of the solution to undetected equipment malfunction. This allows for the possibility of equipment cross-checking itself. We must be wary of common-mode failures however (§2.3.2). If there can be a common cause of two or more redundant systems failing, then the redundancy no longer helps us to isolate and eliminate equipment malfunction.

5.8 Estimates of serious human error

The ideal inspection would ensure that human error and equipment malfunction never occurred, never influenced the outcome, or at least never went undetected. However, this, if conceivable, would probably be infinitely expensive. Even expensive single items such as aeroplanes, bridges, a Channel tunnel, offshore oil platforms or nuclear reactors, might be uneconomic to build and use if we had to make the inspection as good as that. Nevertheless, even though we can make the inspection very good, we should also estimate the chance of a serious human error that may go undetected in order to have confidence that nothing has been overlooked. We have in mind the errors of kind described in §5.5.3.

Some of the methods of estimation are: Delphi technique; fault-tree analysis, hierarchical task analysis; rare-event analysis. There is insufficient space here to describe all these evaluation techniques in great detail. The first two were discussed in Chapter 2, and the statistical aspect of rare events are given in the Appendix. Here we consider the Delphi technique briefly before presenting some results obtained with it.

In the Delphi method, a group of experts is consulted in an iterative way. They are first asked to reply to a questionnaire which contains questions covering the topics, such as the chance of human error occurring in an inspection, in a comprehensive manner. These questionnaires are then analysed and a majority view formed. This majority view on the questions is then sent back to the experts together with another questionnaire covering the same ground as before perhaps with some narrowing down in certain areas. The experts respond again, usually with a greater degree of agreement between the responses. This process can be iterated several times until a concensus view is obtained. This view is taken to be a best estimate of the effect which has been considered. In this way Marshall (1976, 1982) obtained estimates of the reliability of ultrasonic non-destructive testing for finding defects of various sizes in thick section steel plate. Subsequent experimental trials such as the UKAEA Defect Detection Trials have provided evidence that this estimate was surprisingly good. The UKAEA Defect Detection Trials results are reported by Watkins *et al* (1983a,b) and are discussed in more detail in Chapter 6.

Bogie (1982) has given an estimate of the likelihood of a fairly serious human error as about 0.0001 to 0.001. The Delphi technique was used to estimate the likelihood of human errors in the manufacture of a pressure vessel. The results obtained were:

an error of design of 0.0001;
substitution of material or supply, error of 0.001;
weld rod substitution, likelihood of 0.006;
error in ultrasonic examination, including the likelihood of a human error, of 0.02.

Presumably some of these numbers can be related to likelihoods of a human error in purely non-destructive inspection. For example, we could expect a choice of the wrong equipment of about 0.001. The human element in the error of ultrasonic inspection of 0.02 gives an indication of the need for clear procedures and possibly for repeat inspections coupled with a sufficient level of motivation and supervision. Such provisions would, presumably, reduce the effect of human error down to the level of the fairly serious errors given above, that is a rate of 0.0001 to 0.001.

5.8.1   Estimate of rate of serious human errors from common experience

Above, we presented a figure of between 0.0001 and 0.001 for the rate of serious human error. Another way of arriving at this figure is to consider the case of human beings performing a relatively skilful task according to a prescribed code. We will illustrate this with an example taken from a completely different field in order to make the point that human beings make mistakes at a certain rate, almost independent of the nature of the tasks in hand. This gives us a base rate below which it is very difficult to obtain error-free human activity.

Consider driving a car. We may assume that accidents would never happen if everyone drove strictly according to the rules as laid down in the highway code taking due cognisance of prevailing weather and road conditions. Whether or not this is strictly feasible does not affect this argument. Statistics are published on the number of accidents (reported) per kilometre travelled. In 1981, for example (Central Statistical Office 1983), there were 248 276 road traffic accidents resulting in personal injury. Figures are usually given in terms of injuries and deaths per 10^8 kilometres travelled. In 1983 there were 324 840 casualties caused by traffic accidents in 10^8 kilometres travelled. Now we can assume that some of these accidents would not have occurred if everyone had been following the highway code strictly. Some accidents would still occur, for example those caused by trees falling on cars or pedestrians during gales, or those caused by drivers suffering from sudden disablement whilst driving (strokes, heart attacks and so on). For the sake of an order of magnitude calculation such as this let us take it that each category of accident—avoidable and unavoidable—is equally likely, then the number of avoidable accidents becomes 162 420 in 10^8 kilometres or about 2×10^{-3} per kilometre travelled.

We need to relate this to the occurrences of hazards which, if the driver fails to react correctly, will lead to an accident. This will be different in towns and cities to quiet country roads but a guess at a reasonable figure would be a few such hazards per kilometre. Here we are thinking of such things as bends, entrances to fields, road junctions, other road users and so on. We see that at two occurrences per kilometre of such hazards, we would anticipate the chance of human error occurring at about 10^{-3} and note that we have considered only the accidents reported, which will undoubtedly be an underestimate. We have presented an argument taken from a different field which provides us with a basis for believing figures of this order of magnitude which have come out of other scientific studies. The point is that it appears to be very difficult to reduce the incidence of serious human errors below some low level and, whilst this cut-off level may be lower than one in a thousand, it is probably not very much lower.

5.8.2 *Experimental values for human error rates.*

Hagen (1982) reported some work originally performed at Brookhaven National Laboratory which provides another quantification of human error rates (abbreviated to HER in some literature). These estimates were deduced from the interface between human operators and the instrumentation and control components of licensed nuclear power plants. A value of human error rate of 8×10^{-3} was found for fairly serious error—in keeping with the other estimates we have produced so far.

Reporting on the inspection of Defect Detection Trials Plates 1 and 2 at the symposium on the UKAEA Defect Detection Trials, Murgatroyd *et al* (1983) gave one particularly interesting result concerning the distribution of errors, as well as a demonstration of good capability inherent in high sensitivity techniques. The interesting result, from our point of view here, was that the experiments provided experimental evidence of two levels of errors. Errors are frequently classified into hierarchies such as level 1 errors, level 2 errors or errors of degree, errors of kind and so on (Bogie and Beyers 1982, 1984, Coffey 1982). This classification of errors depends on whether the errors occur frequently (level 2, errors of degree etc), or whether they are rare (level 1, errors of kind etc). Usually the rare ones are those with the most disastrous consequences for the reliability of inspection. An example of such an error would be if the inspection were omitted or the results overlooked completely. In Murgatroyd *et al*'s results a frequency distribution of errors in the height of the defects as measured, compared with the actual values found from destructive examination, reveals a normal distribution of errors together with a constant term independent of defect size. In other words, there is a small finite chance, not zero, of getting larger errors on the size of the defect even in tests of this sort carried out under laboratory conditions with a great deal of cross checking by motivated personnel. Another by-product of the UKAEA Defect Detection Trials was an estimate of the chance of human error. Out of 270 defects actually detected by all the teams, one defect was not reported due to an error in transcription of the results. A rough estimate of the likelihood of this simple mistake occurring is therefore $1/270$, that is about 4×10^{-3}, which agrees well with the figures we have been extracting throughout this section.

Such levels of error do not necessarily lead to reservations about the integrity of the structure, as has been pointed out for the pressure vessels of pressurised water reactors (Marshall 1982, Cameron and Temple 1986) provided there is inherent structural safety built in due to good engineering design, correct choice of materials, and careful workmanship during manufacture; this is discussed in §6.9.2.

5.9 Human factors and how to control them

Human beings are sensitive to many external stimuli, and it is therefore wise to be aware of the many factors which can improve or impair their performance when involved in non-destructive inspection. There is a whole field of study, ergonomics, devoted to studying such factors and we shall simply list some results which are available in the references cited.

The performance-shaping factors which are most appropriate in non-destructive inspection are the following (not necessarily in order of importance).

Selection of appropriate personality.   Introverts often perform better at signal detection tasks than extroverts (Gange *et al* 1979, Davies and Hockey 1966).

Design of 'good' inspection procedures.   Wright (1971) found that good layout of inspection procedures can sometimes reduce the error rate in understanding by a factor of about two. Similarly, if transcription of results is required then bad layout can increase the chance of error by a factor of about five (Klemmer 1969).

An instruction to 'take care'.   The exhortation to 'take care' can improve performance by a factor up to six in certain cases (Williams 1969, Herr and Marsh 1978, Hagemaier *et al* 1978).

The good presentation of signals.   In the presentation of signals there is an optimum speed with rate of presentation which changes every one to three seconds (Jenkins 1958, Conrad 1951). The response time for perception of displayed signals is such that at about 0.2 seconds exposure, instrument reading errors can increase by up to a factor of ten (Elkin 1958) compared with exposure time controlled by the inspector.

Use of multiple sensory stimuli for signals.   If a signal is to be reliably detected then it is better to present the stimuli to more than one sense simultaneously. For example a signal which is presented both aurally and visually will be detected about 15% more reliably than a signal which is only presented either visually or aurally (Osborn *et al* 1983).

The avoidance of intrusion of visual noise.   Visual displays that are cluttered were considered by Wolf and Green (1957) who found that detection probabilities of signals in visual clutter can fall by more than an order of magnitude in going from noise dot densities of about 20% of the screen area to coverage of about 60% even with extended viewing time.

In using human operators, who may be required to scan for long periods or over large areas before defect indications are found, there may be some useful knowledge to be gained from other fields of endeavour. For example, radar monitoring is quite closely related to the monitoring carried out by ultrasonic inspectors, in the sense that the information is presented in the same way via a cathode ray screen. Baker (1958) considered the plan position indicator presentation of 'pips' and indicates a mean detection failure probability of about 0.46. For a B-scan presentation, Baker and Boyes (1959) gave a mean probability of failure to detect of around 0.18. As the time on the task increases, so the human operator becomes less alert. The mean failure to detect signals increases from 0.07 to 0.18 over two hours (Mackworth 1950) whilst extending this period to eighteen hours increases the failure probability to about 0.7 (Baker *et al* 1962). Breaking up long stints of inspection into shorter sessions can reduce the probability that mistakes will occur. For example Colquhoun and Edwards (1970) found in a disc discrimination test that the failure probability decreased from 0.6 to about 0.4 by splitting up the test into eight 40 minute sessions.

A review of the published literature relevant to human reliability in complex environments can be found in Embrey (1976), where the conceptual background to the analysis of human reliability data is discussed.

5.10 Properties affecting the performance of the technique

Having designed an inspection procedure based on a technique which is capable of detecting the defects of interest, and having tried out the technique and procedure on some real examples, we may be dissatisfied still to find that the successful detection and classification rate is not as high as we had hoped. This may be because there are factors which were not considered at the design stage, yet which impair the performance. Such factors include the following.

(*a*) Crack transparency.

(*b*) Applied stress tending to close cracks.

(*c*) Some material included in the defect (for which the approved technical term is *crud*) which makes it harder to detect.

(*d*) Poor access (especially for manual operators).

(*e*) Component or defect geometry of an awkward sort.

(*f*) Some component or other property which reduces reliability such as cladding, surface finish, barnacles on the surface, temperature or degree of magnetisation or level of radioactivity.

We should also consider the environment.

(*a*) Is there any background radiation (for x-ray and gamma-ray techniques)?

(*b*) Are there any stray electric fields (for electrochemical and resistivity and eddy current techniques)?

(*c*) Are there stray magnetic fields (for magnetic particles, eddy currents)?

As an example, Zambon (1971) points out that some problems in aeroplanes occur in flight and are correctly diagnosed by flight crew but back at the maintenance depot the fault cannot be made to recur. One reason for this can be the difference in operating conditions; in flight at 33 000 feet the pressure is only 4.4 psi (1 psi = 6.895×10^3 Pa) and the temperature is $-25\,^\circ$C whereas on the ground the pressure is 14.7 psi and the temperature may be as high as $30\,^\circ$C.

5.10.1   Effects of compressive stress

We must consider each of these in a little more detail. Taking them in order, we note that defects such as fatigue cracks can have faces which are close together and which may have many small points of contact. This will have the effect of reducing the ability of techniques such as eddy currents, x-rays, or ultrasonics to detect these defects. In the case of eddy currents and ultrasonics this is because the fields (electric or elastic displacement) can be propagated across the microbridges formed at the contact of asperities, and the effectiveness of the defect in altering the fields measured is reduced. For radiographs, as we saw previously, a small gap between the faces of a defect makes the defect particularly sensitive to orientation effects on the signal. As the applied compressive stress increases, the two faces of the crack will be pressed closer together. The signal will, in general, be reduced for any of the three techniques listed.

For ultrasonic inspection with normal incidence of compression waves on a tight fatigue crack whose faces are under compressive stress, Haines (1980) developed a model in which the flow stress of the material and the properties of the surface, such as root mean square deviation from flatness, and also the applied stress, affect the reflection and transmission coefficients. The result for the transmission coefficient was

$$T = 2\left(2 + \frac{i\omega\pi P_{m}\bar{r}Z_{1}}{SkE}\right)^{-1} \tag{5.5}$$

where ω is the angular frequency of the incident plane wave, P_{m} is the flow stress of the material on each side of the crack, $\bar{r}$ is the mean radius of contact points at the asperities on the surfaces, Z_{1} is the acoustic impedance of the medium, S is the applied stress, k is a constant which depends on the statistics of the surface but always takes a value of magnitude about 2, and E is Young's modulus for the material. Typical values of these constants for steel would be an applied stress S of 300 MPa, a flow stress of

$P_m = 1500$ MPa, $\bar{r}$ of about 20 μm, Z_1 would be about 46.8×10^{-7} Pa s m^{-1} (the obsolete unit of specific acoustic impedance, the Rayl, has been super-seded by the SI unit Pa s m^{-1} where 1 Rayl $= 10$ Pa s m^{-1}), and E is 2.1×10^{11} N m^{-2} and these give a transmission coefficient of $|T| \simeq 0.92$ at 2 MHz and $|T| \simeq 0.67$ at 5 MHz. The modulus of the reflection coefficient, $|R|$, is given, in this case of normal incidence, by

$$|R| = 1 - |T|. \tag{5.6}$$

Therefore, with most ultrasonic inspection techniques, the signal amplitudes for this defect under these conditions would be about $20 \log |R|$, that is 22 and 9.6 dB respectively, below their optimum value. These values are in general agreement with experimental values obtained by Wooldridge and Steel (1980). Temple (1984a) has extended this model to incidence at arbitrary angles and correctly includes mode conversion effects allowing predictions of the amplitude of ultrasonic response of tight fatigue or hydrogen cracks under compressive stress to be made. Typical results for the reduction in signal which occurs from a tight fatigue crack under large compressive stress can be seen in figure 5.4. This shows that even under compressive stresses representing about 75% of the yield stress, the signal is only reduced by a maximum of about 12 or 13 dB compared with the signal from a completely open defect. This is, in fact, very encouraging.

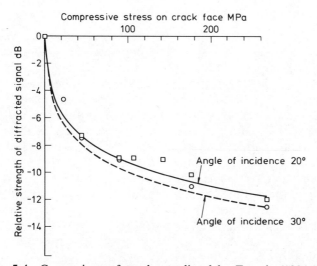

Figure 5.4 Comparison of results predicted by Temple (1984a) and experimental measurements of Whapham *et al* (1984). The theoretical curves are for a surface RMS roughness of 1.5 μm, the flow pressure of the material is 1200 MPa, the frequency of the ultrasonic compression waves is 6 MHz. Experimental results for 20° incidence are denoted by □ and for 30° incidence by ○.

5.10.2   Ingress of alien material into defects

Even if the two faces of the defect were not in intimate contact it may be possible for there to be an ingress of some other material. Examples here could be water, for submerged welds as in offshore oil rigs and pipelines or indeed in any welds in vessels containing water, or other liquids, such as liquid sodium in a fast reactor, or there could be corrosion, such as rust or various oxides. An investigation of the effects of such material included in cracks has been made by Temple (1980, 1981, 1983a). The model used is based on the exact reflection and transmission coefficients for plane waves incident at arbitrary angles on an infinite crack-like defect with parallel faces separated by some specified distance. This model has been extended to incorporate small amounts of roughness on the faces of the defect, but so that the faces of the defect still do not have many points of contact. A typical result (Temple 1983a) is that for a 2 μm wide defect entirely filled with rust the reflection coefficient of a 4 MHz shear wave incident at 45° can be reduced by up to 62 dB by the presence of the rust. If the same defect were filled with water, the reduction in signal amplitude would be about 10 dB compared with an air-filled defect. This is shown in figure 5.5. Depending on the reporting thresholds laid down in the inspection procedure, difficulty might be experienced finding large enough signals to report in either of these cases.

5.10.3   Access and component coverage

Another cause of an unreliable inspection with an otherwise capable technique can be the poor access to the component. This does not occur in automated inspections with kit designed for the particular job unless there is a problem with setting it up, but it can be a considerable problem with manual inspections. This is relevant to any of the common inspection techniques, such as eddy current, radiography, ultrasonics and magnetic particle inspections. One of the most likely occurrences here is for the test equipment to fail to make contact with the component under test over some area.

A solution to this problem can be to use digital storage of the signals obtained from the equipment. Subsequent analysis would then show up if contact had been maintained throughout. This may not be much help if access to the component is not possible (as for example in a nuclear reactor) until some later time when the same problem may arise again. In these circumstances it is beneficial to have some interactive storage of coupling information and automatic warning to the operator if the test equipment ceases to make adequate contact with the component. This warning might range from simple audible indication to more sophisticated displays on cathode ray tubes or to visual display units of position-encoding microcomputers.

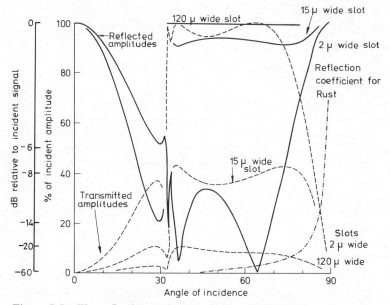

Figure 5.5 The reflection and transmission of 4 MHz shear wave at a water-filled defect as a function of the angle of incidence. Also shown is the reflection coefficient for a 15 μm wide defect containing 'rust' (taken to be Fe_2O_3 with the elastic constants of haematite). Reflection coefficients are shown as full curves and transmission coefficients are shown as broken curves.

5.10.4 Comparison of radiography and ultrasonics for tight cracks

Different physical factors can affect each inspection technique differently. O'Brien *et al* (1975) have given a good account of one particular case involving inspection of the lugs on a rudder actuator fitting on a jet transport aeroplane. Stress-corrosion cracking was found to be the cause of a failure and eddy currents were used for inspection. On several occasions small cracks, about 3 mm long, were detected, but could not be confirmed with dye penetrant inspection. Ultrasonic inspection of a similar lug revealed a stress-corrosion crack 15 mm in length and depth and which could not be found with dye penetrants. Radiography then revealed a flaw about 15 mm long which did not extend to any surface, whereas eddy current inspection revealed a crack only 3 mm long. To understand this discrepancy a section through the lug was taken revealing a crack of area about 200 mm^2. The crack was very tight close to the surface and only about 3 mm long at the surface. Two important conclusions drawn from this case history are the importance of crack tightness and the relative ineffectiveness of radiography for very tight cracks, despite the large loss of specimen thickness and the ideal orientation of the x-ray beam.

5.10.5   Component curvature

Other important points to bear in mind are the geometry of the component and also (as noted in Chapter 4) such properties as the surface finish, temperature, vibration and so on. Workpieces which have curved surfaces, especially if the curvature is fairly tight, might give problems with equipment designed and tested only on flat plates, for example. The results of this might be inadequate coupling or perhaps an error of a completely different type.

As an example of a different type of problem, we note that the ultrasonic time-of-flight diffraction technique relies on timing measurements between pulses which are diffracted by the ends of a defect and, possibly, a signal which travels just underneath the surface of the component. This works very well provided that the velocity of propagation is known, which in general it is. However, on components with curved surfaces, the wave which propagates around the curved surface does not have the usual velocity of propagation equal to that of the bulk wave, nor does it have the velocity of a Rayleigh surface wave. Unless this speed is known the values obtained with the time-of-flight measurements will not be as accurate as is usual with the technique. This particular problem has been solved, and the speed of propagation around cylindrical surfaces for compression or shear waves can be shown to be

$$\frac{v}{c} = [1 + 0.928(k_i a)^{-2/3}]^{-1} \tag{5.7}$$

where v is the required speed of propagation, c is the bulk compression or shear wave speed, k_i is the wavevector of the ultrasonic wave and a is the radius of the concave surface (Peck and Miklowitz 1969, Hurst and Temple 1982). Experimental confirmation of this result was reported by Charlesworth and Temple (1981). This illustrates further that the problem anticipated is not necessarily the one which actually makes the technique fail: one might have assumed, reasonably, that the speed would remain constant, but that experiments would have detected a discrepancy between actual and measured defect position.

5.10.6   Ultrasonic inspection through anisotropic cladding

In the pressure vessel of a pressurised water reactor, the inside surface is covered with two layers of austenitic steel. Because this crystallises with a large grain structure, it is very anisotropic compared with the smaller-grained ferritic steel usually encountered. Thus ultrasonic techniques might have difficulty inspecting beneath this cladding for a variety of reasons. The microstructure of this cladding can be unpredictable and, even if it is well characterised, the particular orientation prevailing might be incompatible with the angled beam specified in the inspection procedures. Wooldridge and Duffy (1982) have given experimental results for the effect of the austenitic cladding structure on the detectability of defects for pulse-echo ultrasonic techniques. A material which is anisotropic for ultrasonic wave propagation may still be inspected with ultrasound, provided the physical behaviour is understood. Also, of course, it may be possible to utilise the other inspection techniques, such as radiography, in these components. The main trouble with reactor components is that the component thickness is considerable (about 250 mm typically) compared with the size of defect which is sought (say 25 mm). The layer of cladding will be of varying thickness due to the way in which it is deposited as a molten layer with subsequent cooling by convection, and the orientation of the grains will have some randomness superimposed on an overall pattern. For ultrasonic inspection this could have had an effect similar to the optical effect produced by the lumpy glass used in bathroom windows.

Fortunately, it has been shown that this is not what happens in practice on real clad material and that, provided the cladding is accounted for either with an empirical correction (Murgatroyd *et al* 1983, Bowker *et al* 1983) or using a theoretical model of the cladding and a calculated correction (Curtis and Hawker 1983, Charlesworth and Temple 1982, Wooldridge *et al* 1982), then accurate results can be obtained with ultrasonic techniques for inspection of clad plate from the clad side.

5.11 A mathematical description of unreliability

When we wish to investigate quantitatively the reliability or unreliability of a given NDT technique we must use statistical concepts. In order to continue the flow of this chapter, mathematical details are reserved for the Appendix.

Consider a non-destructive examination in which some parameter, like the amplitude of a signal on an oscilloscope screen, is compared with that from a reference reflector. Let us call the amplitude from the reference reflector A_c and that from the 'defect' A_s then, according to the inspection procedure rules, we report the defect if $A_s \geqslant A_c$. However, in practice both A_s and A_c will be different each time they are measured, even for the same defect and the same calibration reflector. This variability is common to any physical measurement and is represented by using a distribution about some mean value. A useful (but by no means the only useful) distribution is the normal or Gaussian distribution. In this, the probability of obtaining a value lying between x and $x + dx$ for some parameter x is given by $P(x)dx$ where

$$P(x) = \frac{1}{\sigma\sqrt{(2\pi)}} \exp\left[-\frac{1}{2}\left(\frac{(x-\mu)}{\sigma}\right)^2 \right] \qquad (5.8)$$

where μ is the mean and σ the standard deviation of this distribution. In these discussions, the significance of σ is that it represents the precision with which the measurements can be repeated, a small σ indicating a high degree of repeatability of that inspection. The distributions need not be symmetrical but a normal distribution is representative for the sort of discussion of errors likely in the measurement of a signal amplitude or the position of an extremity of a defect.

Suppose next that both the actual defect signal and that from the calibration reflector are given by this type of distribution. This should often be a satisfactory hypothesis, and is used for illustrative purposes here. We assume also that the two distributions have the same standard deviation σ; this simplifies the analysis as well as being a justifiable approximation in many cases. The means from the defect and calibration experiments are μ_s and μ_c respectively. A defect goes unreported if $A_s < A_c$. In figure 5.6 we show by means of the hatched area the region in which the defect gives a smaller

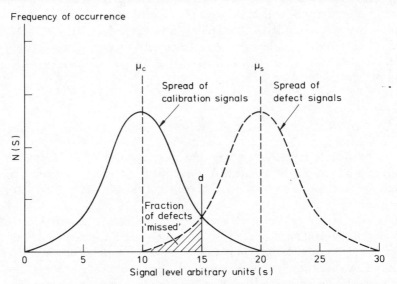

Figure 5.6 The reliability of defect detection with techniques based on the comparison of signal amplitudes. The hatched area denotes the fraction of unacceptable defects which are likely to be missed. This depends on the variation between repeated measurements of both calibration and defect and on the separation of the mean levels $(\mu_c - \mu_s)$.

signal, although the defect is actually bigger than the calibration and ought to be reported. The probability of incorrect reporting in this case is $B(x)$, say, where $B(x)$ is given in the Appendix and where a plot of the function $B(x)$ is given in figure 5.7. The figure has a characteristic S shape. The method we have indicated above, and enlarged in the Appendix, is related to 'signal detection theory' which is used to estimate the failure rate to detect and correctly classify rate and, conversely, the false call rate.

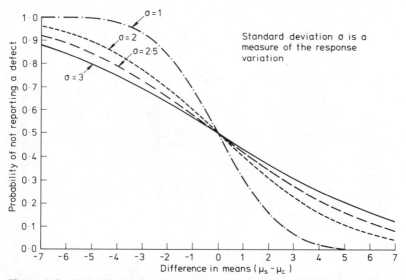

Figure 5.7 Reliability of defect detection assuming that the errors of measurement and calibration are normally distributed as shown in figure 5.6.

5.11.1 Defect sizing errors based on measurement of defect extremities

Another related problem is that of measuring a defect size from two separate measurements of position of the defect ends. This is common practice in most conventional ultrasonic inspection techniques and also in the time-of-flight ultrasonic technique, for example. What is usually important in these inspections is the ability to reject correctly all defects above some specified size. Let us assume, for the present, that the distribution of measured position about a mean at each end of the defect is normally distributed and that the means of the two distributions accurately represent the actual positions of the ends of the defect. Suppose we know that the distributions at each end are Gaussian with means μ_1 and μ_2 respectively and that the standard deviation of both distributions is equal to σ. This could represent the measurement of defect through-wall size by time-of-flight, for

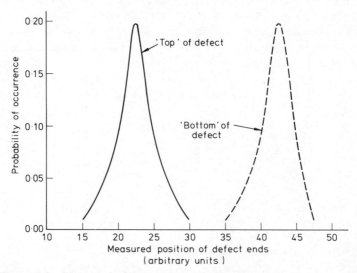

Figure 5.8   Many techniques measure the size of a defect by measuring the position of the two extremities and subtracting the values. There is a probability distribution for the position measurement of each end as depicted schematically in this figure.

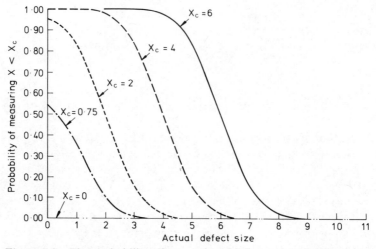

Figure 5.9   The probability of measuring a defect to be of a size x less than some stipulated critical size x_c as a function of the actual size. All units of size are in multiples of $\sigma\sqrt{2}$ where σ is the standard deviation of the measurement.

example, and is illustrated in figure 5.8. Derivation of the results is included in the Appendix, here we simply plot the results. Plots of $1 - P(s \geqslant x_c)$ are given in figure 5.9 for $x_c/\sigma\sqrt{2} = 0, 0.75, 2, 4, 6$ and for $0 \leqslant |(\mu_2 - \mu_1)/\sigma\sqrt{2}| \leqslant 9$.

5.11.2   Defect under-sizing or over-sizing

Another interpretation can be placed on the results in §5.11.1, where previously we took the two means μ_1 and μ_2 to be accurate representations of the actual defect extremities and the parameter x_c to be specified as a critical size, we can also take x_c to represent the actual defect size and μ_1, μ_2 to be experimental values. Then we can associate $P(s \geqslant x_c)$ as the probability of measuring a defect which is bigger than, or equal to, the actual defect size. For example, from figure 5.9 we see that for a real defect size $x_c = 4\sigma\sqrt{2}$ and if the measurements give $|\mu_2 - \mu_1| = 3\sigma\sqrt{2}$ then the likelihood that we measure a defect bigger than the real defect is only about 0.16. An estimate of the chance of measuring $s \equiv x_c$ is given in the Appendix.

5.12   Repeating inspections

To lead into the subject matter of the next chapter, which discusses experimental trials that can be used to determine either the capability or the reliability in practice of inspection techniques, we consider here how an inspection reliability will be improved by repeating the measurements. Suppose we have an inspection with a probability P of detecting a certain class of defects. This value of P may have been measured or may be known from theoretical modelling work. Then if we repeat the inspection we should improve our chance of finding defects of this particular class. On the first inspection we have a probability of finding the defect of P and, therefore, a chance of missing it of $1 - P$ whilst on the second inspection we find another $P(1 - P)$ and miss $(1 - P)^2$ so that the probability of a successful inspection is now $1 - (1 - P)^2$. For n inspections the likelihood of success is

$$1 - (1 - P)^n \tag{5.9}$$

as given by the binomial distribution (see the Appendix). To see what this means in terms of improvement in reliability, take the case where an inspection is only 65% successful, then $P = 0.65$ and after two inspections we shall have improved the reliability to about 88% and if three inspections were permitted we could hope for 96% success.

This all seems very encouraging, and might suggest that even with an appalling technique we could still be reasonably successful if we repeat the inspection several times. However, the above equation applies only if the inspections are independent and the errors random. In practice this could not be achieved by the same inspection team repeating their measurements, since they would probably be influenced in each inspection by what they had discovered during earlier inspections. The improvement in reliability would not necessarily follow by using different teams either, since the teams themselves may have very different abilities. And even though the technique and the procedures were identical the different teams might obtain results of different quality (see discussion of PISC I results by Haines *et al* (1982)).

The probabilities of detecting or not detecting defects which are published as the result of round-robin test-block trials are averages over may defects. In attempting to apply these probabilities to other defects we note that any particular defect could give a consistently small signal because of effects such as material ingress, component geometry, defect geometry and so on. In such cases, the signal returned is not a randomly distributed quantity (at least not with the same mean and standard deviation as that for the defects used in the round-robin test) and no number of repeat inspections will find it with the *average* probability reported from the round-robin trial. If it consistently returns a very small signal, which is below the capability of the technique to detect, then repeating the inspections a large number of times will still leave the probability of detection essentially zero (i.e. if P is almost zero then the expression above departs from zero only for very large n).

In discussing the reliability of inspection we really mean *the reliability of detection and classification of defects in a specified class*. The class might be chosen on the grounds of defect through-wall size or position; defect length; crack thickness or orientation; crack type, such as stress-corrosion, lack of fusion in a weld; or in a host of other ways such as discussed in greater detail in Chapter 3.

5.13 Summary

Any non-destructive testing technique consists of an inspection technique, the actual equipment, and a set of rules or inspection procedure for using the equipment. Human elements enter in several places. In the case of a manual inspection, the equipment is applied to the component by a human being, or, in the case of an automated inspection, by a machine set up by humans. The results whether recorded or not are interpreted in terms of defect indications and characteristics.

The following points were considered in the chapter.

(*a*) The appropriateness of the technique for defects sought.
(*b*) The validation of the inspection procedure.
(*c*) The application of the technique according to the procedure and the

possible limitations from equipment failure, from the ways in which human error could influence the results, from properties of the defect which have not been considered and which might impair the performance of the technique, and from the need to treat experimental results statistically.

In the next chapter we consider the use of round-robin test-block trials in estimating the reliability of non-destructive testing techniques and review the results obtained in some recent studies.

6

Idealised inspection studies

To err is human, but really advanced disasters need a computer

As we have already seen, the overall reliability of an inspection depends on the successful combination of a capable technique and its application in practice without error. If the safety of an aeroplane, a bridge, an offshore structure or a pressure vessel depends on knowing whether there are defects present, and how big they are, then we need to know the probability that all significant defects have been found. Armed with this knowledge we can then make assessments of the expected reliability, or of the life expectancy, of the structure, based on our knowledge of operating conditions and fracture mechanics rules.

Information on the probability that a given technique will detect the defects of concern is useful both in the improvement of a technique or in choosing the best technique out of a range of possibilities. How do we obtain this information? If the technique has been used extensively in the field, and if knowledge of the number of successes and failures is known from, say, destructive examination, or if the number of structure failures which can be attributed to inspection failure is known, then the reliability of the technique can be estimated statistically. However, the information required for this is often not known. The number of destructive examinations may be small, or few structures may have been inspected, or the technique itself may be new. There is no universal solution to this problem. An attempt may be made through the use of test-block trials, in which blocks are examined with the non-destructive testing technique. The blocks are then cut up so that the defects found destructively can be correlated with those found non-destructively.

In mass production, with large numbers of inexpensive items, it is not hard to extract sufficient statistical detail to quantify the reliability of the NDT method. For expensive single components, however, such as the node of an offshore structure, or a reactor pressure vessel, the cost of providing a statistically meaningful number of specimens can be intolerably high. Nevertheless, test-block trials can be set up to verify two distinct aspects of the inspection technique, namely the inherent capability of the technique, and its reliability in practice.

We look next at these two aspects in detail and shall consider specific results of some test-block trials in order to see how such trials should be organised and what may be expected from them.

6.1 Tests of inherent capability

The capability of the technique depends on the physical principles on which it is based. Often it will be possible to calculate some of the key factors from a sound understanding of the physical principles. Examples of this were mentioned in Chapter 4, and include sizing accuracies of ultrasonic techniques which are determined by the wavelength λ; the detection capability of radiographs determined by the component thickness and absorption and the sensitivity of the film; and the penetration depth of eddy currents determined by the frequency and permeability of the material.

These factors should all have been taken into account when the technique was designed to detect, to size correctly and to characterise the defects of concern. During development of the technique, and following the building of a prototype inspection kit, small samples will be made up containing artificial defects to test the capability of the technique. These samples are testing the physics underlying the technique and are thus confirming the inherent capability. Such tests are even better if they can be made on specimens containing real defects which are later examined destructively.

6.2 Idealised inspection studies: test-block trials

There are several difficulties with test-block trials. First, test blocks typically contain only a few defects, often too few for statistical significance. Secondly, realistic defects are difficult to manufacture to accurate tolerances. Thirdly, the cost of destructive examination is high. It is hard to carry out sectioning to the tolerances necessary for comparison with high-resolution NDT techniques. Finally, one might question the realism of implanted defects. Despite these difficulties, test-block trials can still be realistic in the most important aspects and can serve as a useful confirmation of basic capability of techniques.

6.2.1 Confidence limits on the results from test-block trials

The binomial distribution (see the Appendix) is important in inspection reliability. It also underlies sampling theory, and its derivation in the most

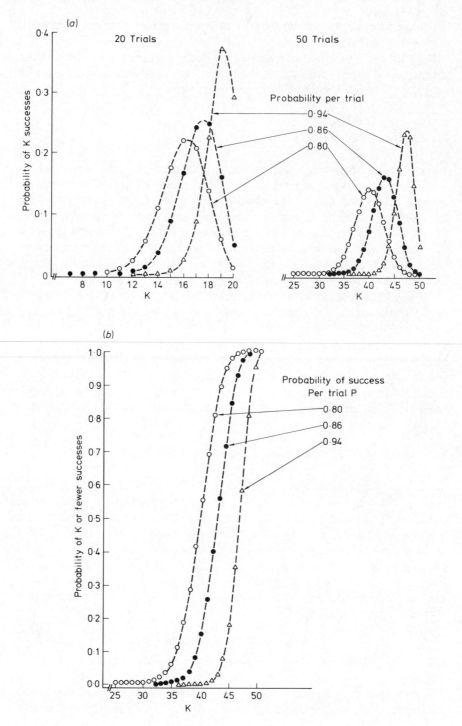

simple way is best demonstrated with an example from that field. Suppose we have a batch of items, some of which are defective, and that the number of defective items is P, the number of perfect items is Q where $P + Q = N$, there being N items in all. If N is very large and we draw out a single item we see that there is a chance P/N that we will have chosen a defective item, and conversely, a chance Q/N that it is perfect. Let us call these probabilities of p and q respectively. Now if we draw out two items we have probabilities that both are defective, both are perfect or there is one of each sort. These probabilities are:

Result	Both defective	One defective one perfect	Both perfect
Probability	p^2	$2pq$	q^2.

Where the terms will be recognised as those of $(p + q)^2$ and, in the more general case, we find that the chance of drawing k defective items in a sample of size n is proportional to $\binom{n}{k}$ which is given by:

$$\binom{n}{k} = \frac{n!}{k!(n - k)!}. \tag{6.1}$$

The binomial distribution thus gives us the chance of exactly k successes in n trials, and p is the probability of success in any one trial. Sequences of events in which successive trials are independent and at each trial the probabilities of the outcomes are constant are called Bernoulli trials. This is a situation occurring often in engineering. In the case here of successful or unsuccessful detection and characterisation, the probability of success or failure remains constant between each trial. Thus,

$$P(k \mid p, n) = \binom{n}{k} p^k (1 - p)^{n-k} \tag{6.2}$$

denotes the probability of obtaining exactly k successes in n independent trials where the probability of success is p on each trial and is plotted in figure 6.1(a). If we want the probability of k or fewer successes in n independent trials we must use the cumulative binomial distribution:

$$Q(k \mid p, n) = \sum_{r=0}^{k} \binom{n}{r} p^r (1 - p)^{n-r} \tag{6.3}$$

Figure 6.1(a) Binomial distribution giving the chance of exactly k successes in n trials with the probability of success on each trial of P. The distribution is defined only for integer values and the broken lines are just to guide the key. (b) Cumulative binomial distribution giving the probability of k or fewer successes in $n = 50$ trials with the probability of success of P in each trial. Note that the distribution is defined only for integer values and that the broken lines are just to guide the eye.

which is plotted in figure 6.1(b). The events are a complete set, that is out of n trials, which all succeed or fail, the result is certain to be one of $[0, 1, \ldots, n]$ successes, i.e. there is unit probability of n or fewer successes

$$Q(n \mid p, n) = 1. \tag{6.4}$$

The probability of more than k successes is given by $1 - Q(k \mid p, n)$.

We can also put confidence limits on the results of the probability of success p given an observation of k successes out of n trials. The value k/n is an unbiased estimate of p and the 95% confidence limits on p are plotted in figure 6.2 for some sample values of p as a function of n. The question is often asked 'How many trials do we need to ensure that our estimate of the reliability of this technique is, say, 95% with a high degree of confidence?' Suppose we wish to obtain this value of 95% reliability for a given defect class and we want this to 95% confidence limits. We need to solve the equations

$$\sum_{r=0}^{k} \binom{n}{r} p_1^r (1 - p_1)^{n-r} = 0.025 \tag{6.5}$$

for p_1 for the upper limit and

$$\sum_{r=k}^{n} \binom{n}{r} p_2^r (1 - p_2)^{n-r} = 0.025 \tag{6.6}$$

for p_2 for the lower bound. We can either stipulate a value for p which we desire to obtain with a given confidence limit and solve for n and k, or we can be given the results of a trial, k successes in n trials say, and estimate $p = k/n$ with confidence limits p_1 and p_2. In either case the equations can be solved by trial and error, using a bisection technique for small values of n. For large values of n we can make use of the approximation of a binomial distribution by a normal distribution (see the Appendix) and obtain, for example, the 95% confidence limits on the value of p from

$$P[\bar{p} - \bar{q} < p < \bar{p} + \bar{q}] \sim 0.95 \tag{6.7}$$

where

$$\bar{q} = 1.96 \left(\frac{\bar{p}(1 - \bar{p})}{n} \right)^{1/2}. \tag{6.8}$$

The value of 1.96 comes from the integral of the normal distribution, i.e. it is the number of standard deviations from the mean which encompasses 95% of the probability on one side of the distribution. For large values of n, Packman *et al* (1976) made use of the Poisson distribution which is satisfactory provided p is either very small, $\leqslant 0.1$, or very large, $\geqslant 0.9$ say. The difference in accuracy between the two distributions is illustrated by Packman *et al* (1976): for $n = 45$ and $k = 43$, then for 95% confidence limits we obtain $\bar{p} = 0.904$, whereas the true value from the binomial distribution would be $\bar{p} = 0.863$.

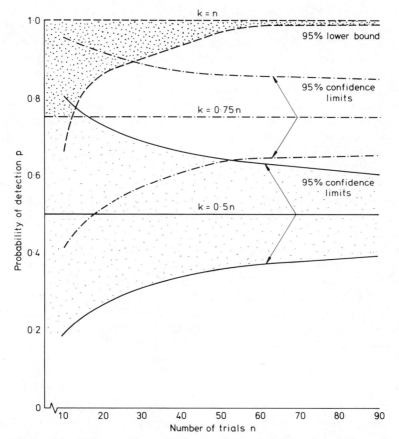

Figure 6.2   95% confidence limits on the probability of detection p estimated by k successes in n trials. Fixed values of k/n are plotted. The lighter stipple indicates the lower and upper 95% confidence limits when the number of successes equals half the number of trials n, and the dark stipple indicates the lower 95% confidence limit for the case when in n trials there are found n successes.

6.2.2   Examples of confidence obtained from a typical test-block result

Figure 6.1 plots the binomial distribution for $n = 20$ and $n = 50$ and for probabilities of success on each test of $p = 0.80$, 0.86, 0.94 and the associated cumulative distribution for $n = 50$. The results are only determined at integer values of k, so the broken lines are of no significance but merely draw the reader's attention to the relevant points for a given value of p. For a test in which there are 15 defects and we are successful in detecting, sizing or correctly classifying only 14 of them, we obtain a best estimate of the reliability of 0.93. Also, we can be about 83% confident that the reliability exceeds 0.80. Another example is a test in which there are 20 trials and we are successful with only 19 of them. The best estimate of the reliability is then 0.95 and we can be about 83% confident that the true reliability exceeds 0.85. In a test with 30 trials and we are successful with only 29 then the best estimate of the reliability is 0.97 and we can be at least 82% confident that the true value exceeds 0.90.

Here we could use the number of trials to mean the number of defects or the number of independent attempts to measure the defect, but we shall use the term to mean separate defects. Similar results for 95% probability of detection with 95% confidence have been given elsewhere (Whittle and Coffey 1981) are requiring 92 successes out of 93 trials. The number of defects required to establish 90% probability of detection with about 80% confidence level is realistic, and shows that test-block exercises could demonstrate such a reliability for a particular class of defects relatively effectively. However, if we were foolish enough to request a 99.5% reliability with 95% confidence we would find it a very exacting task. We would then need to be successful in all of about 600 trials or, if we failed only one test, we would need success in 949 out of 950 trials.

Clearly, there is a moral here for anyone requesting experimental justification of a technique to too high a degree of confidence, unless they can afford considerable expense of time and money! Similarly, if an adequate demonstration of reliable inspection is required over many different defect classes then huge numbers of test blocks will be required.

6.3   An example of the cost of fabrication and sectioning of test blocks

In the UKAEA Defect Detection Trials (Watkins *et al* 1983a,b) there were four specimens, of which three were flat plates and one simulated the geometry of a pressurised water reactor (PWR) nozzle inner radius. The specimens used were of full thickness and of reactor quality, clad as though each was actually part of a PWR vessel. Plate 1 contained 29 deliberately introduced flaws and plate 2 contained 16. The defects in these plates were of planar, crack-like type distributed throughout the entire weld volume of the 250 mm thick steel plates. Plate 3 contained 26 deliberately introduced flaws in the region near to the austenitic cladding whilst specimen 4 contained 20 flaws near the inner radius and extending into the nozzle bore and to the vessel face. Thus there was a total of 91 flaws. This limit on the total number of defects illustrates the above problem with test-block exercises, namely that it can be expensive to obtain a statistically meaningful number of defects, and this was a serious criticism of the PISC I (1979) results by Whittle and Coffey (1981). The point is illustrated in figure 6.3 where the cost in millions of US dollars, in 1984 prices, of manufacturing and sectioning test blocks of reactor-quality steel containing a number of realistic defects is given. Across the bottom of the illustration are given the various combinations of defects and successful detections and characterisations which we have discussed above. We assume that there are about 15 defects in a test block and that the cost of material, fabrication and sectioning of the block is roughly $100,000. The cost of various numbers of defects is shown on the left-hand vertical scale, and the associated reliability scale is shown on the right-hand vertical scale.

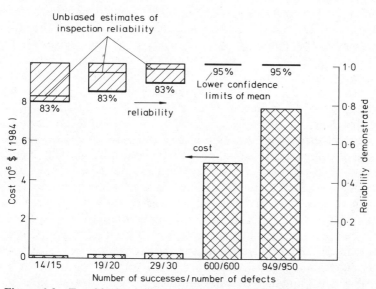

Figure 6.3   Test-block trials: cost of material, fabrication and sectioning. Single hatching gives the unbiased estimate of the reliability of inspection (horizontal line), the bold line gives the lower confidence limits about the estimate and the figures state what the confidence level is. The double hatched regions give the cost in millions of dollars (USA 1984) of the test blocks.

An example will help to explain the remaining information on figure 6.3. For a test-block exercise containing 15 defects in which success is obtained on 14 of them then, as we saw above, the reliability is about 0.93 and we can be about 83% confident that the reliability exceeds 0.80. This is shown as a bold line for the estimate of reliability in the upper part of figure 6.3 with the confidence band shown hatched. In this case we have illustrated that at the 83% confidence level the reliability exceeds 0.8. Moving across the figure to a trial with 30 defects and success in 29 of them, we see how the cost doubles and there is an improvement in the estimated reliability and the confidence levels. At the far right of the figure is illustrated the cost of trying to demonstrate very high levels of reliability.

6.4 Manufacture of test blocks

Not only is it difficult to arrange for sufficient defects to be introduced into test blocks to give results in which there is a high statistical confidence, but it may also be difficult to manufacture the blocks to the exacting standards which are required in order to test techniques of high intrinsic capability.

In table 6.1 the errors, which occurred between the intended defect sizes in plates 1 and 2 of the UKAEA Defect Detection Trials and the minimum defect volumes found destructively, are analysed in terms of their mean and standard deviation for all three dimensions. The Z dimension is the important through-thickness dimension, with X the distance along the butt weld of the plates and Y the coordinate along the surface of the plate perpendicular to the weld. The results show a mean error of 3.8 mm or less, with a standard deviation of 14.7 mm at worst. For the important through-wall size, the mean error was 2.9 mm in plate 1 and only 0.4 mm in plate 2, with standard deviations of 2.7 and 7.2 mm respectively. The correlation coefficients between the intended defect sizes and those found during the destructive examination are given for the first two blocks of the Defect Detection Trials. A correlation of at least 0.89 was achieved for the through-wall extent, which shows that fabrication methods for these thick section steel test blocks achieved a consistently high standard. Figure 6.4 is a histogram of the errors which occurred in the through-wall size of defects inserted into plates 1 and 2. These are errors between the intended sizes of the deliberately implanted defects and those actually manufactured as later determined by destructive examination.

Table 6.1 Comparison between intended defect size and minimum defect volumes found destructively for Defect Detection Trials plates 1 and 2.

Plate	Parameter	Correlation coefficient	Mean error	Standard deviation or error	Maximum undersize	Maximum oversize
1	X size	0.980	3.8	3.7	−8.0	16.0
1	Y size	0.955	0.2	1.3	−5.0	2.0
1	Z size	0.978	2.9	2.7	−9.0	5.5
2	X size	0.766	−1.7	14.7	−47.5	20.5
2	Y size	0.814	2.3	6.7	−7.5	13.0
2	Z size	0.882	0.4	7.2	−7.2	17.0

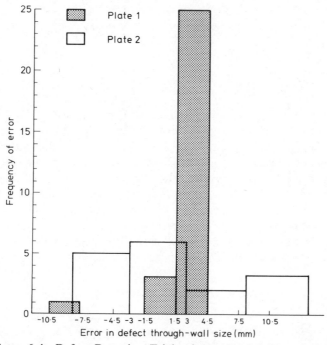

Figure 6.4 Defect Detection Trials plates 1 and 2. Errors in defect through-wall size as manufactured compared with intended size.

6.5   Destructive examination

After making test blocks and inspecting them, destructive examination is necessary. This presents a problem in terms of the engineering ability required to cut up thick section steel plates with the tight tolerances necessary for accurate comparison of the results with sensitive ultrasonic techniques. In the Defect Detection Trials, the plates were cut up into smaller cuboids containing the defects and then these were examined using a combination of very high sensitivity ultrasonics, metallography and further sectioning.

In the destructive examination of the plates, small defects may be associated with the intended defect. These can occur due to imperfections in, say, the welds used to implant defects and a rule must be followed in order to decide whether such associated defects should be included in the destructive results. It is important to ensure that the same rule is followed by the teams reporting on the ultrasonic round-robin results, especially if the comparison between the non-destructive and destructive results is to be made on the basis of simple boxes drawn round defect *'extremities'*. In the results which we present in later sections, we have used the minimum volumes of the defects found during destructive examination except where it is stated explicitly that the comparison is with the extended defects including the defects associated with the implanted coupon welds.

6.6   Realism of test blocks

Another problem with test blocks exercises is that, because of the desire to introduce as many defects as possible within a limited budget, the defect density present in the blocks may be several orders of magnitude greater than would be reasonable in practice. This has at least two possible effects. First, defects may be inserted so that they obscure each other (which would be acceptable if it is likely to occur in practice). Secondly, the reality of an inspection scanning several metres of weld without finding a defect is lost.

This is not to say that test block exercises have no value. On the contrary they play an important role in demonstrating the capability of teams and of techniques to detect, size and characterise defects.

6.6.1   Realistic defects

In Chapter 3 we described the types of defect of concern. Not all these defects are easy to simulate with high fidelity in test-block trials. However, much progress has been made, and now it is possible to simulate such defects as tight fatigue cracks, hydrogen cracks, intergranular stress corrosion, lack of weld fusion, slag, and carbon or copper contamination. Details of the techniques used for defects in thick section steel blocks have been given by Watkins *et al* (1983a,b, 1984).

6.6.2   Environment

One aspect of test-block trials which does not usually mimic real inspections is the environmental conditions. Test blocks are usually examined under laboratory conditions without the difficulty of access, without extremes of temperature, and without need for protection against radioactivity and other problems which may occur in the actual inspection. These factors can be controlled and simulated if necessary, but this generally adds still more to the cost of the trials. Moreover, such extra features may add little of scientific value to the trials, though they may still be useful demonstrations for improving public acceptability of the usefulness and relevance of the non-destructive test under trial.

Sometimes techniques developed under laboratory conditions are totally unsuitable for field use because of some environmental condition which was not considered. For example, the equipment might need to operate near a welding machine and may have been conceived without due regard to the electrical interference from it. In such cases, test-block trials in an appropriate environment would be an essential part of the proving process for the equipment.

6.6.3   Design of trials

To derive the maximum benefit from a test-block exercise, careful thought must clearly be given to objectives. The experiments must be carefully designed, bearing in mind their relevance to the structural integrity of, say, the final component. To this end it is vitally important to extract the maximum amount of information on the test itself as well as on the specific results stemming from it. In most cases, this analysis should be undertaken before any subsequent test-block exercises are considered.

6.7   Results of test-block exercises

In looking at the results of a test-block exercise we should draw attention to two particular facets. We have called these *interpretation*, and *interpolation*, and we discuss them in separate sections below.

6.7.1   Interpretation of results from test-block trials: bounding boxes

If a test-block trial is to serve a useful purpose, the results must be analysed carefully. All too often the results are analysed in a very rudimentary way which fails to extract the true value of the tests. The problem is usually one of comparing results from a destructive examination with these from the non-destructive examination. In early test-block trials the *detection* of defects was all that mattered. Boxes were drawn round the actual defects and those reported from the non-destructive test. If these boxes overlapped then a defect had been successfully detected, and the technique was judged on these successes. Let us consider this. If the boxes reported by the non-destructive test are very big, then they will almost certainly overlap those found destructively, and a high *detection* score will result. But suppose the reported boxes were about the right size, yet displaced relative to the results from the destructive examination. Then under what conditions could we say fairly that successful *detection* has occurred? The answer is determined by the aim of the inspection. If the inspection is used to determine whether a component is suitable for some specific, normal, or postulated abnormal, operating conditions then there will also be a criterion under which the component will be unsuitable. Often this will be expressed in terms of a maximum crack size which can be tolerated in the component. In certain cases where unacceptable defects are found, repair of the component will be permitted rather than scrapping it. In these cases, a defect would still be considered as correctly *detected* if, during excavation of the material in the vicinity of the reported flaw, one requiring repair was actually found. The method of such grinding away, or excavation of material prior to repair, determines the tolerance with which an unacceptable defect needs to be found during test block trials. In general, though, it is better not to be content with the overlap of simple bounding boxes as the criterion for *detection* in test block trials.

6.7.2   Other measures of accuracy of defect detection and location

A better approach to the use of bounding boxes would be to associate a probability of detection based on the distribution of the geometrical distance between the centre of gravity of the defect as determined destructively and that reported from the non-destructive test. Alternatively, projections of the area of the defect could be made onto three mutually perpendicular planes, and the success score could be derived from the ratios of these areas as determined non-destructively and destructively. In other words, we should look at the mean and standard deviation of the distribution in location and sizing errors of defects at least. The better techniques are then those with the smallest mean errors and the smallest standard deviations.

6.7.3   Use of the results of test-block trials

Not all dimensions of a defect are equally important in determining the severity of flaws and this should be taken into account in the reporting of the results. For example, it is the through-wall dimension of crack-like flaws in thick section structures which is of most importance in determining the structural integrity.

If results are poor, so that large mean errors or large standard deviations occur, we should return to basics and check the physics, the type of defects inserted (are they realistic and representative?), and the operation of the equipment during the test. We should even be prepared to redesign the instruments to handle the situation better in the future.

6.8   Interpolation of results

Provided that the test-block exercise has been well designed, the results should give information on the usefulness of the inspection technique in detecting, sizing and characterising defects of particular types and sizes. The size range used should be that with the greatest significance for ensuring the integrity of the component in service. The results refer necessarily to the discrete values of defect size and type in the samples tested. However, in practice the technique will be applied to a whole spectrum of defect types and sizes. To infer the probability of correct characterisation of defects over this whole spectrum from the measurements made at discrete values we must perform interpolation (and possibly extrapolation, although it must always be remembered that extrapolation is often a divergent process which will yield unrealistic values if carried too far), entailing fitting an appropriate function to the discrete results. Once this fit has been achieved, we have information which, essentially, does two things. It calibrates the system over a wide range, and it allows inference of the actual defect population in a real test from the measured defect distribution.

6.8.1 Making use of test-block results

This leads us to the sort of information required in a probabilistic fracture mechanics assessment of integrity of a whole structure. The usual way of carrying out the interpolation is to fit a function, $B(x)$, say, to the data to represent the chance that an unacceptable defect with a range of parameters x will be allowed to remain in the component after inspection. In principle, $B(x)$ should be the probability that the unacceptable defect is left in the component following inspection and repair, but this is a further stage which is beyond the scope of this text.

It is important to note that we have chosen to represent this reliability of inspection function as a function of a vector of parameters x. Some of the parameters which x represents are: the dimension of each of three orthogonal directions; the orientation of the defect in three dimensions; the location of the defect in three dimensions and proximity to other defects, component boundaries or changes in cross section. Other parameters which might be included could be the surface finish of the component, the surface roughness of the defect, the nature of the defect such as included material, or the ambient conditions around the defect of stress, or other fields.

This impressive list of parameters should be compared with what is often recorded about defects in practice. It is quite common to find studies in which the defects found are reported as a simple number, i.e. '3 per metre of weld', without even a definition of size in one dimension! In practice, of course, the integrity of the component is our main aim and this will be determined by only a few of the above list of parameters. For cracks it will often be sufficient to consider only one parameter such as length or through-wall extent. The function $B(x)$ which enters a probabilistic assessment of integrity will then actually be $B(x)$ with x simply the length or through-wall extent of cracks. However, we must remember that it is not always the same parameter which governs structural integrity that is crucial to the successful detection and sizing of that defect.

6.8.2   Estimates of reliability based on modelling

Often little is known about the function $B(x)$ which characterises the reliability of the non-destructive test. It can be estimated by fitting appropriate functions to results of test-block exercises or other comparisons between non-destructive and destructive tests or it could be derived from first principles if sufficient were known about the physics of the technique and the defects of concern. This latter approach, of modelling the response of a variety of defects to an inspection, is a fruitful one. Examples of the approach to ultrasonic examination can be found in Coffey *et al* (1982) for conventional ultrasonic examination of thick section steel, in Temple (1984b) and Ogilvy and Temple (1983) for the ultrasonic time-of-flight diffraction technique, in Haines *et al* (1982) for conventional inspection of defects, and in Thompson and Achenbach (1985) for the ultrasonic inspection of branched stress-corrosion cracks with either dB drop or time-of-flight techniques. An example of the modelling approach to estimating the reliability of defect detection and sizing for radiographic inspection is given in the article by Segal *et al* (1977).

Coffey (1978) used Kirchhoff theory to calculate defect responses from rough defects and found a Rayleigh distribution of echo amplitude. As a typical result, at 50% DAC, with a 70° probe, Coffey calculated the probability of detection for a crack, with a rough surface such that the response was down by 26 dB at normal incidence compared with a perfect reflector. He found a probability of detection as low as 0.35, but at a sensitivity of 20% DAC this increases to 0.85. Haines *et al* (1982) showed that, for defects 25 by 125 mm in 200 mm thick pressurised water reactor vessel seam weld, a probability of detection of 0.99 is obtained for the pulse-echo corner effect at 2 MHz. This calculation allows for up to 20° of tilt. The tandem technique for the same defects gives a probability of detection of 0.98 at 1 MHz. For defects 5 by 25 mm the probability of detection values are 0.81 for the pulse-echo corner effect at 2 MHz for defects tilted by 20° and a probability of detection of 0.95 for the tandem technique at 1 MHz. All these data apply for a sensitivity of 3 mm DGS. (DGS is a system of compensating for the variation in signal strength with distance of the defect from the transducer. It is rather like DAC (see §5.3.2) but using a flat-bottomed calibration reflector instead of a side-drilled hole.)

6.9   Probabilistic fracture mechanics

In §2.4.1 we introduced the concept of component replacement strategies based on probabilistic methods. Engineering structures are designed to allow for some possible failures. Sometimes a component can fail without affecting the overall integrity but whether this is the case or not it is always recognised that failure is possible. Failure occurs if loads exceed some critical value, or if material strength or toughness is too low for the applied load, or if the material contains cracks or stress concentrators which magnify the load locally. As we have seen in Chapter 3, defects can grow during service for a variety of reasons and environment can age materials leading to reducing toughness with time. To cover these possibilities an approach used in the past has been to apply factors of safety in the design of structures. These safety factors are usually grossly over-conservative and increase the cost of structures enormously. A modern approach is to design a structure to accommodate a crack of predetermined size without failure. In this way one can do away with the 'thoroughly bogus "factors of safety"' (Gordon 1978, p. 103). If the material properties are considered variable, or a spectrum of loads is possible with different frequencies, or if the reliability of non-destructive inspection is considered an important part of ensuring the integrity of a structure, then a useful approach is to use probabilistic fracture mechanics. With a probabilistic fracture mechanics approach, possible variations in the factors listed can be used to predict a failure rate for the structure. This predicted failure rate tells us how many structures of similar construction could be expected to fail if we built very many of them. In other words it predicts the mean failure rate. It would be possible in principle to carry out the calculations to obtain the likely fluctuations in failure rate about this mean value. However, in practice the information required concerning the distributions of the input variables, such as material toughness, initial crack size distributions and so on, is only known sufficiently well to estimate the mean failure rate, and effort is concentrated on finding this and its sensitivity to the input parameters.

The mean failure rate of an ensemble of similar structures is predicted but in practice there will often only be one structure, or at best a few, built. This failure rate can be interpreted as the likelihood of failure of an individual structure out of the ensemble. The term failure rate is the technical term used in the probabilistic fracture mechanics literature and is interpreted in the likelihood sense in this section.

6.9.1   Mathematical expressions for the possibility of inspection failure

In the probabilistic assessments of component integrity a variety of forms for $B(x)$ have been assumed. In work on ultrasonic inspection of thick ferritic steel components, Marshall (1982) proposed the following form for the chance of leaving an unacceptable defect in the vessel

$$B(x) = \varepsilon + (1 - \varepsilon)e^{-\mu x} \tag{6.9}$$

as a function of x taken to be through-wall extent only of the defect. This function was based on a limited number of actual data combined with a

Delphi questionnaire approach to experts. The values obtained were $\varepsilon = 0.005$ and $\mu = 0.113$ (mm^{-1}). This expression is interesting because there is an asymptote ε below which the chance of leaving an unacceptable defect in the component never falls. This asymptote is to be thought of as arising from factors beyond the scope of the non-destructive inspection technique. An example of the source of ε would be human error, such as omitting the inspection or being distracted before reporting a significant result. The Marshall $B(x)$ was based on information available prior to 1976 and is a function only of the through-wall extent of a crack. The results of the PISC I trials (PISC 1979) suggested that the reliability of ultrasonic inspection in thick section steel as given by the Marshall $B(x)$ was optimistic, but the intervening years have seen substantial improvements in inspection capabilities.

It is clear that, for ultrasonic inspection, long defects should be easier to detect than short ones of the same through-wall extent. This was demonstrated in the UKAEA Defect Detection Trials. Cameron and Temple (1985) have modelled this to produce a function $B(x)$ which takes account of the length of the defect and the ultrasonic scan pitch to produce a new effective constant μ. The results, for a scan pitch of 15 mm or less, and for a defect through-wall extent x measured in millimetres, give

$$B(x) = 0.005 + 5.53e^{-0.37x} \qquad (6.10)$$

for defect aspect ratios of $1/3$ and

$$B(x) = 0.001 + 10.3e^{-0.57x} \qquad (6.11)$$

for defect aspect ratios of $1/10$, provided $x \geqslant x_a$ where x_a is an acceptable defect size.

Harris (1982) used a function denoted $P_{ND}(x)$ for our $B(x)$ given by

$$P_{ND}(x) = \varepsilon + \frac{(1-\varepsilon)}{2} \, \text{erfc}(v \ln x/x^*) \qquad (6.12)$$

where x^* is the crack through-wall dimension giving a 50% chance of correct detection and classification. Harris concluded from several sets of data that the values of x^* varied from about 2.5 to 9 mm, with $\varepsilon \sim 0.005$, and v in the range 1.33 to 3.

A model for $B(x)$ encompassing many variables, several inspection techniques Q and human variability was proposed by Temple (1982):

$$B_Q(x) = \frac{1 + \exp[-\sum_i \alpha_i^*]}{1 + \exp[\sum_i \alpha_i(x - x_i^*)]} \qquad (6.13)$$

where there are i parameters in x. The physical significance of the parameters α_i is that they determine the range of x values over which the transition from acceptable to unacceptable defects occurs, that is the sharpness of the boundary between acceptance and rejection. The parameters x_i^* are then the values at which the probability of correct rejection reaches 0.5

if there were only that parameter present. When there are several techniques Q and the kth is used n_k times on a population of defects which can be split into N classes, then the total population of defects missed is B given by (Temple 1982)

$$B = \sum_{i=1}^{N} c_i \prod_{k=1}^{M} [B_{Q_i}]^{n_k} \tag{6.14}$$

where c_i is the proportion of the total defect population which is in class i.

Temple's model is rather similar to a model developed in the study of the reliability of non-destructive testing methods in the US Airforce and called 'the log odds' model. The model is so called because logarithms of probabilities are used in linear regression fits to the data but, from the results obtained, it could easily have been called long odds! The log odds model gives a probability of detection POD(x) given by

$$\text{POD}(x) = \frac{\exp(\alpha + \beta \ln x)}{1 + \exp(\alpha + \beta \ln x)} \tag{6.15}$$

(Berens and Hovey 1982, 1983), and hence a chance of leaving the defect in the vessel $B(x)$ given by $1 - \text{POD}(x)$:

$$B(x) = \frac{1}{1 + \exp(\alpha + \beta \ln x)}. \tag{6.16}$$

The US Airforce tested this model by transporting sections of retired aircraft and other specimens to several airforce bases and these specimens were examined by representative personnel, using various non-destructive tests in a typical inspection environment. After the teams had inspected the specimens, destructive examination was used to determine the actual population of defects. This programme is often referred to as the 'Have-cracks-will-travel' or 'Have cracks' for short, and the data generated by it as the 'Have cracks' data. Both the 'Have cracks' data and simulated data from a Monte Carlo sampling programme were examined using the log odds model and reported in Berens and Hovey (1983).

6.9.2 Relevance of probabilistic fracture mechanics to required reliability of inspection

The probabilistic fracture mechanics approach can be used to study the effects that unreliable NDT will have on the integrity of the component under assumptions about the distributions of material toughness, about the initial defect densities, about the likely crack growth rates under the various environmental conditions—temperature, stress, chemical attack and so on.

In §§6.8 and 6.9 an expression for the chance of leaving a defect of size x in the component after manufacture and inspection was discussed. This function, denoted $B(x)$, is used in probabilistic fracture mechanics assessments of the predicted failure rate of the pressure vessel of a pressurised water reactor (Marshall 1982, Cameron 1984, Cameron and Temple 1986). Such calculations have been carried out for the pressure vessel of a pressurised water reactor (Cameron and Temple 1984) and the results are shown in figure 6.5. Based on the assumptions of well known distributions of material toughness and crack growth rates, and on rather less well known distributions of initial defects, the calculations show how the predicted vessel failure rate varies as a function of the parameter μ in the equation for $B(x)$. We recall that large values of μ are appropriate for techniques which distinguish easily between significant and insignificant cracks. On figure 6.5 the log of the predicted failure rate (per vessel year) is plotted as a function of $\log(\mu)$, for μ in mm^{-1}. The failure rate plotted is that due to crack initiation during a large loss of coolant accident (LOCA). The frequency of the LOCA is assumed to be 10^{-3} per reactor year. For large values of μ the predicted vessel failure rate is dominated by the asymptote ε of equation (6.9). This situation corresponds to a non-destructive test which is intrinsically capable of detecting and correctly classifying every

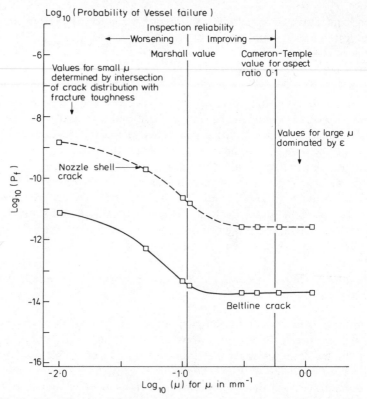

Figure 6.5 The implications of structural integrity requirements on levels of non-destructive testing reliability: an example taken from the pressure vessel of a pressurised water reactor.

significant defect but which is limited by other events, such as human error. At small values of μ the predicted failure rate depends entirely on the overlap of the distribution of material toughness with the initial defect distribution, i.e. NDT is having essentially no effect on the failure rate.

In figure 6.5, the asymptote is fixed at 10^{-3}. Cameron and Temple also considered the effect of varying ε and concluded that it is necessary to increase μ and decrease ε at the same time to achieve the optimum performance. As in the case of figure 6.5, the effect on the failure rate soon saturates if only one parameter is changed. From such an analysis it is possible to work back from desired levels of structural integrity to the level of NDT that is required, provided the distributions of material properties, applied loads and the likely initial defect population are known.

6.10   Role of test-block exercises in estimating inspection reliability

The chance that an unacceptable defect may be left in the component even after non-destructive testing depends on many parameters of the defect and the operating environment. In a test-block trial it is clearly impossible to expect to test them all. This can be seen as follows. If we have N parameters, and we propose to test p values of each, then we shall need N^p specimens. For even modest statistical significance of the results, we might choose p to be as low as five, and then count up our earlier list of important defect parameters. These comprised three linear dimensions, three location coordinates, three orientation parameters, surface finish, defect roughness, and something to characterise the material included in the defect, say density only. This gives twelve parameters or 248 832 tests! Even restricting ourselves to three parameters with five values of each would still require 243 tests. Hence it is essential that attention is restricted to the most important parameters. The most important parameters are those with the largest influence on the reliability of the inspection technique, that is those with the greatest effect on the spread of the results. As noted earlier, this spread is measured by the standard deviation of the results. In talking of standard deviations, we imagine the sort of distribution in results which follows some smooth distribution, such as occurs in the normal carrying out of any experimental measurement. However, as discussed in §5.5.3, there is another sort of error which may not be distributed according to nice smooth distributions. In this category could be really serious errors which have nothing to with the inspection technique itself. An example would be if the inspector, in preparing a final report, lost the record of a serious defect and so did not include it. This is not an error which should occur very often but its consequence could be very serious. An example of such an error, as mentioned earlier, was an error of copying results from one data sheet to a report in the Defect Detection Trials resulting in an estimate of the likelihood of this occurring as 1/274, that is with a probability of around 0.004.

Ways of evaluating the performance of the same team inspecting the same set of defects under different conditions are difficult to devise, since familiarity with the test blocks will affect the operator's judgment on subsequent examinations. To overcome this would need many different, but apparently identical, blocks, which could be mounted in several different orientations. Because of this, the main emphasis in practice on previous round-robin test-block trials has been on determination of capability rather than reliability. Estimates of reliability come from combinations of experiments to test specific points, such as those described in §§5.6.1, 5.6.3, 5.8 and 5.9. These estimates can then be combined in a fault or event tree analysis of the inspection itself to yield its reliability. The use of fault or event trees is discussed briefly in Chapter 2.

6.11   Alternatives to test-block exercises

There are alternatives to test-block exercises. Traditionally, one might do nothing at all to test a new technique, and simply hope for the best. Another, far more sensible approach, would be to combine limited experimental validations with theoretical modelling. Any theoretical models developed must themselves be shown to agree with experimental results over a useful range of validity. If this can be done, modelling provides assurance that the capability of the technique is maintained over a range of parameter values without the cost of reproducing all the results experimentally.

Neither of the two alternatives mentioned can deal with the reliability of the technique in practice. Another option which allows this is to develop a simulator of the inspection. Simulations of the complex operations involved in flying aeroplanes or controlling nuclear reactors are now common, and it would seem possible to also apply the same principles to inspection. Such a simulator would be of value in at least three ways.

(a) It would enable an assessment of the reliability of a given ultrasonic technique to be established with a large number of defects, thus giving statistically significant results with a high degree of confidence;

(b) it would provide a method for testing the competence of inspectors;

(c) it would both simplify and extend the training of operators. For instance, training in the analysis and interpretation of ultrasonic data is an important requirement. A simulator could be employed for this purpose, using information gathered on realistic flaws.

The simulator would also allow two problems to be avoided. One is that, as we have already seen, test blocks are expensive to fabricate and the number of flaws that can be included is restricted. Another is that it is desirable to have precise knowledge of flaw parameters, and this is not always available in the detail necessary. In conventional tests such information is gained by destructive examination which ensures that the test cannot be repeated. A simulator can use the information many times, for it is the signals, not the samples, which need to be preserved. A computer based inspection simulator might be used as a means of assessing reliability by repeated measurement under controlled conditions. It would also allow more fundamental studies of the ergonomic aspects of inspection, and could provide data for input into the models developed for a reliability programme. Simulated experimental conditions could be varied to determine the sensitivity of proposed procedures to such parameters as flaw type, coupling and noise levels.

As envisaged, an inspection simulator would probably work along these lines. A specimen is made up of the correct geometry and surface finish, but containing no defects of a size or type of the sort which would be sought by the technique under trial. This specimen is mounted in an appropriate setting and environment. For example, if access would be restricted in practice, then access would be similarly restricted during the test. The inspector or trainee then proceeds exactly as if he or she were making a real test. Normal flaw detection equipment is employed, except that the probe *position* is continuously monitored. Coupling between the probe and the specimen is also continuously monitored. These determine the simulated defect signals which the inspector sees. Data from the position and coupling monitoring are fed to a computer; as the probe is scanned, flaw signals from the computer are fed into the flaw detector at preselected probe positions, and mimic what would have appeared with the pre-chosen defect. These signals can be a composite of experimental measurements and theoretical modelling work but are envisaged mainly as being digitised A-scans from real defects. Such signals can be arbitrarily disguised to accommodate the effect of noise and coupling variations, say. The inspector responds to the signals as if they were genuine, and records defect indications as he or she thinks fit. Correlations can then be made between his or her recorded indications and the actual positions stored in the data bank to judge the success of the inspection.

6.12 Ultrasonic inspection of thick section steel components: PISC I and its problems

Over the years there have been a number of test-block trials aimed at determining the efficacy of various non-destructive testing techniques. Our aim here, in §§6.12 and 6.13, is to present a list of some of these trials and comment on what in our view are the most important conclusions. For full details the original references should be consulted. We note that, in general, the results of each test present a snapshot view of the capability existing at the time of the test. One would expect the techniques to be evolving and improving, though this may not always be apparent from the results of successive test-block trials, largely because of the organisation of the trials themselves. One should bear in mind that the results of the trial are a convolution of the intrinsic capability of the non-destructive test with the procedure laid down by the trial and with the subsequent rules for comparing non-destructive results with destructive ones.

The inspection of thick section steel, which is important in ensuring the integrity of the pressure vessels of certain nuclear reactors, or of offshore structures, has been the subject of several round-robin trials. Chockie (1978) reported on one of the earliest round-robin trials of ultrasonic capability in thick section steel blocks, and this was followed up by the PISC I trials (1979). A subsequent PISC II trial has been carried out with preliminary results reported in August 1985 (CEC/OECD 1985). Definitive results of PISC II had to await further destructive examination and were available in October 1986 (CEC/OECD 1986).

The PISC I trials involved 34 teams from 10 European countries. The inspection procedure was basically that laid down by the American Society of Mechanical Engineers (ASME XI 1974). This was the first test of its kind to be completed on thick section steels so the results were of great interest, but there was substantial criticism of the trial itself and of the presentation of the results. It is instructive to note the criticisms of this trial so that future trials may be more successful.

6.12.1   PISC I plates

The PISC I trials used three of the plates from the programme of inspection organised by the Pressure Vessel Research Committee (PVRC) in the United States. The teams examined the three thick-section, low-alloy steel plates in which there were artificially induced weld defects. Two plates, intended to be flat, were formed by welding sections of SA302(B) steel plates together. One plate was rather curved, indicating that the correct heat treatment was not applied to the finished weld.

6.12.2   PISC I plate material

The material used was representative of the older material used in reactors but unrepresentative of more modern steels. This resulted in many ultrasonic reflectors other than the crack-like defects intended. There were many inclusions, such as MnS, SiO_2, Al_2O_3, although no complete chemical analysis of the plates was carried out.

6.12.3   PISC I destructive examination

The destructive examination revealed large amounts of porosity near the weld region suggesting, perhaps, that better welding techniques could have been used. Some 500 defects exceeding a sphere of radius 1 mm were found during destructive examination of the three weldments, and the proximity rules laid down for associating neighbouring defects (ASME 1974) would have meant that three quarters of the length of one weld would have been considered as a single large crack! Of these 500 defects only 37 were included in the comparison of the destructive and non-destructive results, and of these only 6 were of such a size as to be unacceptable in a real vessel weld.

6.12.4   PISC I defects and confusion with procedures

Some of the defects were inserted close to the edges of the plate, so that inspection of them was impaired. This would not happen in a real weld, since there are no such plate boundaries. The procedures laid down for the inspectors to follow were ambiguous in places, leading to confusion. For example, the coordinate system had been poorly defined and results were often reported incorrectly. Bench marks on the specimen were also misinterpreted.

6.12.5   Value of PISC I trial

Despite these problems of organisation, which in hindsight could have been improved, the test did provide a useful and relevant test of the inspection procedure. The results showed that all the teams detected both cracks which were larger than 50 mm in through-wall dimension. From this we can infer that the upper limit to the chance of missing defects of 50 mm through-wall size is 0.07 at 99% confidence. Teams were allowed to use advanced techniques as well as the conventional procedures, and these performed significantly better. The original procedure had specified a reporting level of 50% DAC, and even this procedure performed much better if the reporting sensitivity was increased to 20% DAC.

6.13   The Defect Detection Trials organised by the UKAEA

Another criticism of the PISC I trials was that the plates were not clad as would be appropriate in a real reactor. This restriction was eliminated in the Defect Detection Trials organised by the UKAEA. The organisation of the

Defect Detection Trials and the plates used have been discussed by Watkins *et al* (1983a,b, 1984). The trials were organised in 1980 to provide information for the public enquiry into the Central Electricity Generating Board's plan to build a pressurised water reactor at Sizewell in Suffolk. There were four specimens, of which three were flat plates and the fourth simulated the complex geometry of the nozzle inner radius of a pressurised water reactor. Initial reporting of the results was in terms of detection only. Successful detection implied that the box bounding the defect as reported by the non-destructive test was within some tolerance of the actual box discovered during partial destructive examination. Complete destructive examination was not attempted, instead the plates were sectioned into smaller and smaller cuboids with high-sensitivity inspection carried out from all faces of the cuboids at each stage until the defect could be located to within a millimetre.

6.13.1   The UKAEA Defect Detection Trial plates and summary of results

The specimens used were of full thickness (typically 250 mm) and clad as though each were actually part of a pressurised water reactor. Plate 1 contained 29 deliberately introduced flaws and plate 2 contained 16. These defects were of planar, crack-like type distributed throughout the entire weld volume. Plate 3 contained 26 deliberately introduced flaws in the region near to the austenitic cladding and plate 4, the nozzle inner radius, contained 20 flaws near the inner radius and extending into the nozzle bore.

In plates 1 and 2, all seven teams detected all the intentional defects during inspection from the clad face and some additional defects were also found (Watkins *et al* 1983b, Watkins *et al* 1984, Lock *et al* 1983). From the unclad face, which is less important since in-service inspection would normally be carried out from the clad side, the results were not as good, with three teams failing to report two defects. In plates 3 and 4, all planar defects were detected, but one defect was not reported due to an error in transcription. For the non-planar defects, one team out of three reported all the defects, whereas two teams did not report the full extent of a slag line, and one team failed to report the reheat cracking in plate 3 (Watkins *et al* 1984).

6.13.2   Manufacture of test blocks for the Defect Detection Trials

Detection is not the final aim of non-destructive testing since, in general, we need information on the size and character of the defect. The results of the Defect Detection Trials were analysed in terms of sizing accuracy by Temple (1985b). Although Temple's discussion is mainly aimed at the time-of-flight diffraction approach, results for sizing accuracy by all the teams are presented based on the information published following the trials. From this work it became clear that test blocks can be accurately and consistently manufactured to a specification. For example, correlation coefficients between intended defect through-wall size and that found following destructive examination are found to be at least 0.89 for plates 1 and 2. Although the sample considered is limited to 45 defects, this does indicate a satisfactory level of manufacturing capability for deliberately implanted flaws which is necessary if test-block trials are to be designed efficiently and economically. Such a degree of correlation is a necessary pre-requisite if test blocks are to be used as validation tools for inspection teams. Destructive examination of the blocks is difficult to the tolerances now necessary to test high-sensitivity techniques accurately. The destructive results used in this work are believed to be accurate to better than ± 1 mm.

6.13.3 Rules for reporting defects

Another conclusion drawn from the Defect Detection Trials was that the rules for reporting defects (e.g. as ASME boxes) should be unambiguous and should be identical for the ultrasonic and destructive examinations. This is particularly important when combining satellite indications or for ir-regular or complicated defects. For fragmented or complicated defects the comparison of simple bounding boxes as deduced from both ultrasonic and destructive measurements is unsatisfactory and other methods of com-parison should be sought. Detailed comparisons of characterisations of the defect would be more useful. This is required if the full value of com-parisons between different techniques is to be realised. Defects once found and characterised in real inspections are considered as acceptable or unacceptable depending on rules laid down by fracture mechanics on the basis of bounding rectangles. However, much detailed analysis has to be carried out in order to reach the stage of drawing these bounding rectangles on which the final sentencing of the defects depends. There is no doubt that, as fracture mechanics itself becomes more sophisticated and the safety margins can be safely reduced from any existing intrinsic conservatism, then the emphasis will be placed more and more on the capability of NDT to characterise the defect. By developing the ability to characterise defects as thoroughly as possible the NDT community will anticipate this trend and may be able to influence any future developments in the rules governing acceptable or unacceptable defects. In particular, there remains a real need for both NDT and fracture tests on the same samples (see Chapter 3).

6.13.4 Sizing accuracy obtained in the Defect Detection Trials

The Defect Detection Trials were analysed in terms of the capability of measurement of defect through-wall size. The mean error in sizing the through-wall extent of crack-like defects was calculated together with the associated standard deviations by comparing the ultrasonic results which were reported with those obtained during a partial destructive examination of four test blocks. The minimum defect volumes were used in the comparison. A statistical test applied to the results grouped as a single defect class of through-wall extent between 10 and 50 mm suggested that the errors in size measurement are not necessarily normally distributed (Temple 1985b). The mean and standard deviation of the error in size alone are not, therefore, always the best measures to use in inferring the actual defect size from the measured values. The correlation coefficient was calculated as well and was used as a rapid indicator of how well the ultrasonic results cluster around a straight line relationship between the actual defect size and that measured. Best estimates of the actual defect sizes that would be inferred from measured values by linear regression were also given for three teams and for three hypothetical values of measured defect through-wall size.

6.13.5   Errors in defect size measurement obtained in the Defect Detection Trials

The key results obtained from the defect detection exercise for through-wall sizing of defects are, in summary, that the average error, averaged over all teams inspecting plates 1 and 2 together, including inspection from both sides of the plates, was 7.9 mm with a standard deviation of 16.5 mm. Associated with these results is a correlation coefficient between measured and actual defect sizes of about 0.6

Whilst the results averaged over all teams are discouraging, it proved possible to do rather better than the average. Some individual team results were considerably better than the average. The mean errors in the measured through-wall size of the defects found by three teams in the Defect Detec-

Table 6.2   Means and standard deviations of the error in ultrasonic measurement of defect through-wall size for the three best of the seven teams in the Defect Detection Trials on plate 1. The ultrasonic results are compared with the minimum defect volumes found during the partial destructive examination.

Team	Side	Mean error in z-size (mm)	Standard deviation (mm)	Maximum undersize (mm)	Maximum oversize (mm)
A	clad	− 1.4	2.5	− 7.0	4.0
A	unclad	− 1.3	2.0	− 8.0	3.0
B	clad	− 0.8	5.9	− 15.0	20.0
B	unclad	− 1.9	3.4	− 12.0	7.5
C	clad	− 1.3	2.3	− 7.0	2.5
C	unclad	− 1.5	2.0	− 8.0	1.0

tion Trials are given in tables 6.2 and 6.3 for plates 1 and 2 respectively, and are shown on figure 6.6 together with the results on plates 3 and 4. For our purposes here it is unnecessary to distinguish the individual teams and so they are referred to simply as teams A, B and C. Further details of the actual teams and their original results, together with an extended version of the following discussion can be found in Temple (1984c, 1985b).

Table 6.3 Means and standard deviations of the error in ultrasonic measurement of defect through-wall size for the three best of the seven teams in the defect detection exercise on plate 2. The ultrasonic results are compared with the minimum defect volumes found during the partial destructive examination.

Team	Side	Mean error in z-size (mm)	Standard deviation (mm)	Maximum undersize (mm)	Maximum oversize (mm)
A	clad	1.6	8.6	− 18.0	22.0
A	unclad	2.1	6.9	− 7.0	17.0
B	clad	5.1	14.4	− 18.5	43.0
B	unclad	11.6	13.0	− 6.5	42.0
C	clad	4.8	11.3	− 6.5	33.5
C	unclad	2.3	3.3	− 2.5	9.0

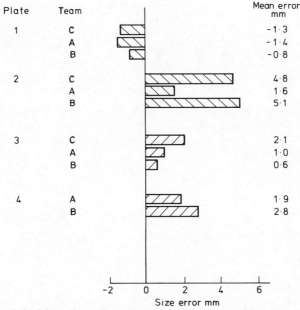

Figure 6.6 Mean through-wall size error with inspection from the clad side of all four plates in the Defect Detection Trials.

The three smallest mean errors of sizing reported in the inspection of plate 1 from the clad side were (Temple 1984c, 1985b) -1.3, -1.4 and -0.8 mm with standard deviations of 2.3, 2.5 and 5.9 mm respectively. These represent good sizing accuracy for a component 250 mm thick covered with a layer of anisotropic cladding in which defects of concern would probably be in the region of 10% of the wall thickness.

In the inspection of plate 2 from the clad side the three best results obtained were mean errors of 1.6, 4.8 and 5.1 mm with standard deviations of 8.6, 11.3 and 14.4 mm respectively. As well as the mean errors, the correlation coefficients between actual and measured defect through-wall size were reported by Temple (1984c, 1985b). These correlation coefficients are given in table 6.4.

In plate 1 these correlation coefficients (0.91 to 0.98) and those in plate 2 (0.67 to 0.98), reflect a consistent and accurate correspondence between the measured and actual defect through-wall dimensions. These results are depicted in figure 6.6

Table 6.4 Correlation coefficients for defect through-wall size, as measured ultrasonically by the three best of the seven teams in Defect Detection Trial plates 1 and 2 and compared with the minimum defect volumes found during destructive examination.

Plate	Side	A	B	C
1	clad	0.984	0.913	0.984
1	unclad	0.988	0.966	0.989
2	clad	0.896	0.800	0.723
2	unclad	0.943	0.674	0.977

Table 6.5 Mean and standard deviation of the error in sizing the through-wall dimension of defects in plates 3 and 4 of the Defect Detection Trial.

Plate	Side	D	E	F	All teams
Mean error (mm)					
3	clad	1.0	0.6	2.1	1.2
4	clad	1.9	2.8		2.4
Standard deviation (mm)					
3	clad	2.6	2.0	4.9	3.4
4	clad	1.4	6.1		4.4
Correlation coefficient					
3	clad	0.946	0.965	0.737	0.886
4	clad	0.985	0.470		0.756

We note that all the teams performing well on through-wall size measurement were using some experimental observations that were not purely amplitude based but incorporated information such as the time of flight of the signals.

The results for plates 3 and 4 are summarised in table 6.5 for all teams who participated in the inspection. These results, and the correlation coefficients for defect through-wall size as measured ultrasonically by the three teams who inspected plate 3 and the two teams who inspected plate 4, are calculated relative to the minimum defect volumes identified during the destructive examination.

The results for all three teams on plate 3 gave a mean error of 1.2 mm with standard deviation of 3.4 mm. Note that these three teams are not identical with those labelled A, B, and C for the inspection of plates 1 and 2. The individual team results for mean errors between 0.6 and 2.1 mm with standard deviations between 2.0 and 4.9 mm are satisfactory. Only two teams inspected the nozzle inner radius specimen, plate 4. The results were mean sizing errors of 1.9 and 2.8 mm with standard deviations of 1.4 and 6.1 mm respectively.

6.14   Further analysis of the results of the Defect Detection Trials

We note that any experimental measurement, such as defect sizing, carries with it an experimental error. Techniques which exhibit a small spread in experimental measurement are to be preferred to those with larger spreads. In practice, when deciding on the size of defects which could be of critical size, this error of measurement would be added on to the value obtained in order to be conservative. The analysis discussed in the previous section together with the correlation and regression analysis which is discussed next could be applied to any similar set of test-block results. We use the Defect Detection Trials' results simply because their analysis has already been published (Temple 1984c, 1985b).

6.14.1   Correlation and regression

One way of improving the reliability of defect detection and sizing with any non-destructive technique is to learn from previous errors. Test-block trials play an important part in this process since the destructive examination can furnish a lot of information about defects, both those satisfactorily detected and sized and those not. This information can be used in a variety of ways to improve future performance. At the level of defect detection it may simply be a case of isolating a human error of, say, having the information recorded indicating a defect but failing to report it. In more complicated cases, especially where an attempt has been made to size the defect, then the detailed geometrical shape or metallurgical structure of the defect might yield clues as to why a particularly good or poor result has been obtained. In field work, however, the detailed nature of the defect is unlikely to be known a priori and the process of determining the defect properties will be iterative. Often the result of a non-destructive test will be a model—conceptual, mathematical, or in some cases an actual model—of the defect. This serves to concentrate the thought processes either into refinement or acceptance of the model as a plausible defect associated with the experimental evidence. If some a priori information is available, and it often is in the form of knowledge about the most likely defects, then we may be able to incorporate this knowledge in our assessment of the experimental evidence. In the case of defect sizing, if we have used the particular NDT technique on many occasions in which the results have been compared with those obtained destructively, then we will have a good idea whether the technique has a zero mean error or consistently under-sizes or over-sizes defects. It would be natural to make use of this information by applying a correction factor—this is the basis of calibration of any equipment. Suppose we have a graph of true defect size plotted against measured size for a range of defects, and that these points are scattered about some straight line (it does not have to be straight but this will make the discussion easier). Then the best fit line through these points yields the appropriate calibration. There are many ways of defining *best fits* of which one way is to use linear regression of the actual defect size on the measured size. Such analysis is readily carried out by standard software packages on computers of any size, from microcomputers to mainframes. The result is a correlation coefficient between the actual and measured values together with an equation for the calibration line.

The correlation coefficient gives a measure of the clustering of the results about a straight line relationship between the actual defect size and the measured value. The smaller the mean and standard deviations of the errors between the measured value and the true value become then the nearer the correlation coefficient becomes to 1.0. Regression of the actual defect size on the ultrasonic measurements was outlined in Temple (1984c, 1985b) based on the results of the Defect Detection Trials. Again the individual identities of the teams are unimportant to the argument and the labels A, B and C are used. These labels correspond to those used in tables 6.2 and 6.3. Results using regression analysis are given for the estimate of actual defect size for three hypothetical measured values obtained by three teams selected from the seven possible by choosing those with the smallest mean errors. The regression analysis results are given in table 6.6 for hypothetical seam weld inspections, with defects resembling those in either plate 1 or plate 2.

Table 6.6

Team	Defects resembling those in plate number	Measured defect size (mm)	Inferred defect size (mm)
A	1	10.0	10.2
B			14.1
C			16.5
A	1	30.0	31.5
B			30.8
C			31.0
A	1	50.0	52.8
B			47.4
C			45.5
A	2	10.0	15.1
B			18.2
C			15.2
A	2	30.0	28.3
B			27.6
C			27.2
A	2	50.0	41.5
B			37.1
C			39.1

Table 6.7

Team	Defects resembling those in plate	Measured defect size (mm)	Inferred defect size (mm)
D	3	5.0	2.7
E			4.5
F			2.5
D	3	10.0	8.2
E			9.2
F			7.0
D	3	20.0	19.1
E			18.6
F			15.9

The results for plate 3 were a correlation coefficient between the ultrasonic measurement of defect through wall-size and the value, found during destructive examination, of about 0.89 for the results of all teams combined. In plate 4 one team achieved a correlation coefficient between ultrasonically measured through-wall size and that determined during destructive examination of 0.99. In a hypothetical inspection for sub-cladding cracks, in which defects of through-wall extent 5, 10 and 20 mm were measured, then the estimates of actual defect size are given in table 6.7.

In a hypothetical inspection of a nozzle, in which measured values of defect through-wall size were found to be 5, 10 and 20 mm then estimates of actual defect through-wall size are given in table 6.8 based on regression of the results for plate 4.

Table 6.8

Team	Defects resembling those in plate	Measured defect size (mm)	Inferred defect size (mm)
D	4	5.0	3.9
E			5.8
D	4	10.0	8.3
E			8.8
D	4	20.0	17.1
E			14.8

6.14.2   Relationship of Defect Detection Trials to vessel integrity

Two teams demonstrated a level of defect detection and sizing which would be sufficient to ensure the integrity of a PWR vessel to currently desired standards. These were the only two teams to complete an inspection of all four plates. Over the first two flat plates, at least three teams achieved the currently desired standard. We note that the defects used in the Defect Detection Trials cover the range of defects which contribute most to the predicted failure rate of vessels—that is of through-wall extent between 10 and about 50 mm (Cameron 1984)—and that it is therefore valid to include results on all four plates in an analysis of the sizing capability of the different techniques.

Repeating the inspections, repeating the setting-up procedure or checking the reporting of defects, are possible methods of reducing the likelihood that any failures to detect or correctly classify significant defects will affect the overall integrity of the structure. Provided the repetitions are independent and do not suffer failure from a common cause, a significant improvement could be obtained, but only if an adequately capable technique is used. Two independent inspections could reduce the asymptote on $B(x)$ to 2.5×10^{-5} but in practice this improvement would probably not be achieved due to the difficulty of obtaining independent inspections which do not suffer failure from some common cause. Nevertheless, it would appear that some useful improvement may follow from repeating inspections with equipment and techniques currently available. There is a difference between using a multiplicity of probes that are moved by one and the same scanning arm and repeating an inspection since, according to our claim, the equipment failures will not seriously affect inspection but the main source of error could be the incorrect setting up of equipment or the reporting of defects. Thus it is the setting-up stage, with its associated risk of human error, which may govern the entire reliability of the inspection and which might benefit from being repeated. This comment also applies to the reporting of defects.

6.14.3 Caveats concerning the results of the Defect Detection Trials

The results analysed are a snapshot of a continually evolving technology. Since the defect detection exercise, improvements have been made to some of the equipment on which these results are based, and subsequent results may be better. Other limitations on the results derive from the management of the trials. The German teams carried out their inspections to a limited cash cost, thereby restricting the amount of time available for detailed analysis; the in-service equipment employed by the French teams was not designed to cope with up to 29 defect indications in about a quarter of a cubic metre (Watkins et al 1983b). The Harwell equipment was not designed to size defects less than 5 mm from the interface between the ferritic base material and the cladding (Gardner and Hudson 1983) and the Risley equipment used on plate 4 contained a systematic error arising from the complicated nozzle geometry (Poulter et al 1982). In the case where defect signals contain indications of satellites, decisions must be taken on how these signals are to be combined and this may be a subjective judgment. An interesting question would be whether the teams would make the same decisions based on their recorded data if it were presented to them again. As evidence of this problem of judgment concerning the reporting of defects with satellites, the correlation results have been given (Temple 1985b) for both the minimum volumes found destructively and for the extended destructive examination which included the implant weld defects. Many of the results improved when the ultrasonic results were compared with the extended defects indicating that the rules for combining defect indications need to be made clearer.

6.15   PVRC—PISC I and other ultrasonic trials in thick section steel

Other test-block trials concerned with ultrasonic detection and sizing of defects in thick section steel are discussed in Chockie (1978). He analyses the PVRC programme, of which PISC I was a European offshoot. The PISC I results are covered comprehensively in the reports (PISC 1979) which are important because they demonstrate how relatively poor was ultrasonic inspection of nuclear reactor pressure vessels at that time, at least if carried out strictly in accordance with the regulations. Since that time, the ASME code for these inspections has been amended, so that ultrasonic inspection has been dramatically improved. Additional techniques, such as time-of-flight diffraction (developed at Harwell), have shown their potential for increased accuracy of sizing of defects. These methods have value too for initial detection, so that use of such advanced techniques will become increasingly common.

6.15.1   Welding Institute tests

Jessop *et al* (1979a,b) reported on a study carried out jointly by the Welding Institute, Harwell, and the Central Electricity Generating Board on size measurement and characterisation of weld defects by ultrasonic testing. This work included a study of the ultrasonic detection and measurement of fatigue cracks in mild steel, and is of special interest because fatigue cracks can be very tight, and may possibly transmit some ultrasound through the defect. Such transparency would reduce the signals reflected or diffracted, and could, therefore, impair either conventional pulse-echo and tandem-type inspections or newer methods like time-of-flight diffraction. Obscuration methods would also be impaired because the defect would cast less of a shadow.

Recent modelling work by Temple (1985b) has shown that the effect on time-of-flight diffraction signals would be of the order of 10 dB at most even if the fatigue crack is under compressive stress equal to three quarters of the yield stress. This result, which depends on the ultrasonic frequency and on the scale of roughness as well as on the material properties of the steel, is valid for frequencies of about 5 MHz, with fine-scale roughness of contact of the two faces of the defect of about 1 μm. For a fixed frequency, the diffracted and specularly reflected signals increase with increasing roughness on the defect faces. For a fixed roughness, the signals also increase with increasing frequency. Experimental demonstration of these properties has been made by several workers, for example Arakawa (1983), Whapham *et al* (1984) and Wooldridge (1979).

The Welding Institute test (Jessop (1979a,b), Cameron *et al* (1981) and discussed by Coffey (1982)) gives the accuracy of a variety of ultrasonic techniques for sizing defects. With conventional manual operation using 20 dB drop and maximum amplitude techniques at 4 MHz with 28 measurements for defects between 1.5 and 30 mm high a mean and standard deviation of size were found to be −1.0 and 3.1 mm respectively. Jig-assisted scanning reduced this variation to −0.7 mm for the mean and 2.4 mm for the standard deviation. This result is for 45°, 60° and 70° probes with 36 measurements. If only the 45° probes are used, even with jig-assisted scanning the mean and standard deviation become −1.2 and 3.4 mm respectively. For the Harwell time-of-flight measurements mean and standard deviations of +0.5 and 1.8 mm respectively were found.

6.15.2 Work outside Europe and the USA

Some of the PVRC and PISC II test blocks were also inspected in Japan, with manual and automated ultrasonic inspections and with automated eddy current inspections. Preliminary results were reported by Ando et al (1982). Aaltio and Kauppinen (1982) have reported on a Finnish study of the reliability of defect detection and sizing. Detection and sizing by ultrasonic and radiographic techniques were compared in butt welds of plates and pipes, nozzle welds, and T-joints in mild steel. The thicknesses of the steel sections were in the range 2–20 mm. The results of the ultrasonic inspections could have been better, with detection probability getting above 80% only for defects with length greater than about 30 mm. Aaltio and Kauppinen considered that the radiographic results showed that that technique was unsatisfactory for crack-like defects and lack of fusion.

6.15.3 Previous studies

Early, yet still useful, references to reliability of inspections can be found in several sources. Thus Packman et al (1975) discuss defect detection in welded structures. Adamenko et al (1979) evaluated the detectability of real flaws using radiographic inspection of welded joints, in work which superseded an earlier Russian investigation by Seminov et al (1973). Rozina and Yablonik (1975) considered some reliability indices of non-destructive testing and the methods of increasing reliability. Their suggestion is to increase the false call rate until the desired probability of detection is achieved. This approach was also suggested by Blistanova et al (1977).

6.15.4 Results obtained in ultrasonic tests

In addition to the results already discussed in §§6.12–6.15, some of the more interesting results to emerge from the studies of ultrasonic reliability are summarised below.

Lecomte and Launay (1979) found, at 95% confidence limits, that there was a 4–5 dB error in the ultrasonic signal from defects less than 150 mm deep and a 6–7 dB error for defects deeper than this. This variation was found to be due to the search unit and the operator, and the results were confirmed by analysis of calibration data taken in 250 tests with 25 operators in shop conditions, where the error was found to exceed 6 dB. Herr and Marsh (1978) observed with a single delta scan of surface breaking cracks with an area normal to the ultrasonic beam of 0.039 square inches, a probability of detection of about 0.85 at the 95% confidence level. Johnson *et al* (1979) used the PVRC data to estimate probability of detection. The PVRC procedure used was the ASME XI procedure, 1974 edition. Data from specimen 201 for defects ranging from 0.05 to 0.17 inches in height gives a probability of detection of 0.985. With confidence limits of 95% this gives a value of probability of detection of 0.937. They believe this value is conservative.

Marshall (1982) in the assessment of the integrity of pressurised water reactor pressure vessels, gave targets for the reliability of non-destructive testing which would be required for a British pressurised water reactor using the best available techniques and procedures. A probability of detection of 0.5 for defects 6 mm high and 0.95 for defects larger than 25 mm high is required. From the results of the UKAEA Defect Detection Trials discussed above, we see that the techniques do indeed have this capability, and from the results of Haines' analysis we see that the current procedures can attain this degree of reliability in practice.

6.15.5 *Comparison of ultrasonic, radiographic and dye penetrant trials*

Magistralli *et al* (1978) considered the effectiveness of ultrasonic, x-radiography and dye penetrant inspections for the detection of cracks in aluminium alloys relevant to aerospace uses. The calibration function of Temple (equation 6.13), when least squares fitted to the results given by Magistralli, yield the values of α and x^* given in table 6.9. The curves are also plotted in figures 6.7 and 6.8 for defect through-wall extent and length respectively.

Table 6.9

Technique	Depth, $\alpha(\text{mm}^{-1})$	Depth, $x^*(\text{mm})$	Length, $\alpha\ (\text{mm}^{-1})$	Length, $x^*(\text{mm})$
Ultrasonics	1.6	0.027	0.44	3.5
X-ray	3.6	0.86	0.18	8.2
Dye penetrants	2.1	0.048	0.63	2.9

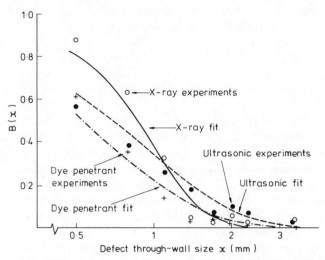

Figure 6.7 $B(x)$, the chance of leaving a defect of through-wall extent x mm in the component after non-destructive inspection. Data points from Magistralli *et al* (1978), curves from fits by Temple (1982).

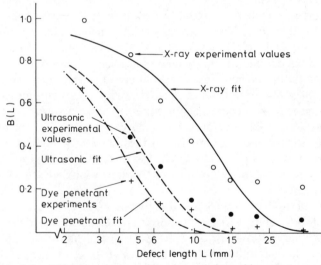

Figure 6.8 $B(L)$, the chance of leaving a defect of length L mm in the component after non-destructive inspection. Data points from Magistralli *et al* (1978), curves from Temple (1982).

In figures 6.7 and 6.8 the circles represent the unbiased estimates of Magistralli *et al* whilst the crosses indicate the upper 95% confidence limits on their results for the chance of missing a defect of given size.

Other work of relevance to the aerospace industry has been carried out and reported by Comassar (1978), Yee *et al* (1976) and Rummel and Rathke

(1974), but the latter two references are now rather old and one expects the effectiveness of the techniques to have improved.

6.16   The future of round-robin tests

Test-block exercises are useful snapshots of the capability of teams using inspection techniques to detect, size and, possibly, to characterise the defects of concern to the structural integrity of components. It is important to know what the test-block trials are intended to demonstrate and to design them accordingly. It is equally important to analyse the results of any test-block exercise as fully as possible to provide quantitative information on the techniques *or* on the trials themselves. Only in this way can the full benefit of such trials be realised and subsequent tests improved.

In our opinion, the thrust of future research should be on the reliable application of techniques which have already proved their capability, coupled with advances in characterisation of defects which have been correctly detected and sized.

In figure 6.9, based on the work of Haines *et al* (1982), we have plotted various ways in which the reliability of any non-destructive testing technique can be improved. The frequency with which signals are observed is plotted schematically as a function of peak signal amplitude. In the top part of the figure the overlap of the signal amplitudes with a reporting threshold is shown. Note that a reporting threshold could simply be a useful way of reducing a large number of data collected on-line by an automated scanner to that which is to be processed subsequently. As the spread in the signal amplitudes is reduced (i.e. as the standard deviation σ is reduced) the reliability of the technique improves as the hatched area, indicating that the number of times the signal falls below the threshold is reduced. The same effect can be seen in the second part of the figure for the case in which the reporting threshold is reduced.

Another method of improving the reliability is to improve the mean signal amplitude relative to the threshold at the same time as reducing the variability. This is shown in the third part of the figure. It can often be achieved by understanding the physics of the inspection and, possibly, by redesigning the inspection technique if it has not already been optimised.

Finally, in the lowest part of the figure, the distribution of signal amplitudes has been truncated. This can be obtained to some degree in practice by validating the testing teams and techniques so that improperly trained operators are not allowed to carry out inspections. This is the licensing solution and could involve the simulator discussed in §6.11. Clearly the optimum solution depends on the various costs of implementing each part, but it will, in general, rely on all three contributions.

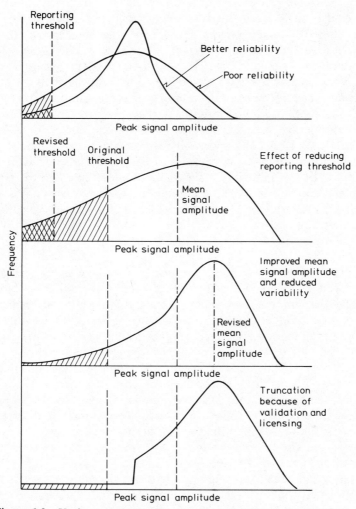

Figure 6.9 Various methods of improving non-destructive reliability.
The hatched areas represent defects which are missed.

6.17 Inspection of other systems

The examples in this chapter have nearly all referred to the detection and
sizing of cracks in thick section steel appropriate for highly stressed steel
structures. The principal reason for this is that, when one attempts to
understand systematically and to quantify the reliability of inspection and
its effects on component reliability, much of the relevant published work is
in the area of structural integrity. We now show that the ideas and
approaches in this specific area have much wider validity.

As we have noted elsewhere in the book, structural integrity is determined by the number of defects present and the stress levels experienced. Cracks are emphasised because they are the most dangerous defects in a stressed structure of homogeneous material (see Chapter 3). The bias towards steel occurs because many stressed engineering components have, in recent times, been made of steel. However, any homogeneous material such as any metal, wood, concrete (to a limited extent), plastics, glass or ceramics behaves in much the same way as steel provided we make allowances for differences in critical crack sizes. The approach to determining and quantifying the reliability of inspection, as we have discussed it in this chapter, is essentially independent of the material as shown by our examples drawn from the aerospace industries, in which aluminium alloys are very important.

Much of what we have discussed in this chapter has direct parallels in many other non-destructive testing applications. In this section we shall briefly indicate how the ideas discussed earlier carry over to these other areas.

Our first comments were on test-block trials. There are difficulties with such trials which would be true of any material including biological materials: the limited number of defects, the cost of specimens, the difficulty of manufacturing controlled defects, and of their subsequent destructive examination, their level of realism and the number of defects per unit volume. The solution requires one to be clear about what it is most important to test. This means knowing which defects are most important: the types of cracks in metals, bone or in wood under tension; the mechanisms of deterioration, such as delamination along interfaces in many biological materials such as ivory and teeth. Tests would need to be made on a range of flawed samples for any non-destructive testing technique and, if the testing environment were important, then these tests would need to be carried out in representative environments.

Modelling can be used to help to understand any non-destructive inspection technique. The technique might involve medical ultrasound, nuclear magnetic resonance imaging in brain scans, or many other more mundane applications. Results from modelling or from limited experiments on test samples will benefit from interpolation. The interpolating functions suggested in §6.9.1 can be used for any technique measuring any parameter of any material, although better functions will exist in some specific cases.

The probabilistic assessment of the structural integrity is a concept which can easily be carried over to other fields. The essence of the approach is simply that there are a number of physical properties which control the structural integrity (examples might be the material strength, the initial defect distribution, the likelihood of defects developing during service due to environmental effects, or the probability that unacceptable defects may be left in the structure following non-destructive inspection). Because these properties are statistically variable quantities, they are treated in a probabilistic way. This is a very broad approach with applications ranging from military wargames to medical efficacy. Suppose we try to devise procedures for assessing the success of a medical diagnosis. We consider several varibles to be direct parallels of those used in structural integrity and these are compared in table 6.10.

Table 6.10

Structural integrity variable	Corresponding medical variable
Material strength	Resistance to infection if disease not diagnosed
Initial defect population	The initial fraction of the population carrying the disease
The likelihood of defects developing during service	The likelihood of vectors to transport the disease
The likelihood of a component being inspected	The likelihood of patient being seen by a doctor
The probability that a significant defect might be left in the structure	The probability of a doctor failing to diagnose the disease or treat it correctly

The outcome of calculations along these lines would be a predicted disease rate for a given population; or such a model could be used to assess sensitivity to changes in medical practice or of more effective tests.

The idea of creating a simulator of ultrasonic inspection has direct parallels, of course, in any field of inspection. The principal difficulties are those of collecting the initial database, of monitoring the necessary degrees of freedom of the inspection tool, and of replaying signals with sufficient fidelity and speed. These problems would need to be overcome in different ways in each case, but the benefits accruing would be common to any technique.

Finally, we note that the criticism of test-block trials discussed in §§6.2.1, 6.4, 6.5, 6.6 and 6.12 would apply to any non-destructive inspection carried out on a series of samples. Clearly one should make sure representative material is used; there should be a balance between the number of non-critical defects and critical defects; and confusion over inspection procedures should be avoided. The maximum benefit is derived from a thorough analysis of the results, rather than mere bland assertions that '...*the technique satisfactorily detects defects of type...*'.

6.18 Summary

For expensive single items it is not possible to obtain the reliability of inspection from the statistics of component failures, or from sampling, as can be achieved on a production line of relatively inexpensive items. One way of estimating capability and (to a lesser degree) the reliability of inspection techniques is the use of round-robin test-block trials. It is difficult to make sufficient of these test blocks for high levels of statistical confidence in the outcome of such trials. Yet provided information is sought on a limited range of defect properties, acceptable confidence is possible. The cost of manufacture and sectioning of test blocks is such that it is very important to design the test carefully and to carry out a thorough analysis of the non-destructive results, preferably with extensive comparison with destructive tests. We have reviewed results for ultrasonic inspections obtained in several recent round-robin test-block exercises as examples of the problems involved and of the practical value of the results.

Computer simulation of inspection is a possible way of avoiding the problem of the destruction of test blocks. It allows the training of inspectors on standard data, and may even permit validation and licensing of teams. It is a field we expect to see expanding rapidly, especially if the promise of the fifth-generation computers is fulfilled.

Note added in proof

The results of PISC II (CEC/OECD 1986) showed that the defect character is the most important factor affecting detection and characterisation with ultrasonic techniques. Planar cracks with sharp edges are always more difficult to detect and evaluate than volumetric defects. The evaluation of defects seems to depend more on factors like location and size than does their detection, with undersizing occurring on average more frequently than oversizing for rejectable defects. As a result of this series of test-block trials, changes to the ASME code were recommended. If accepted, these will reduce the reporting threshold to 20% DAC and will introduce a $70°$ compression-wave probe for near-surface inspection.

Appendix

A.1 Statistics and probability

It is widely believed that those aspects of statistics which are not common-sense are subtle deception. Neither extreme is sufficient for the applications in this book, so we bring together some of the ideas and expressions in this Appendix. No proofs will be given.

First, we ask when does one use statistics?

(*a*) As a measure of error in experiment, or of the precision with which parts are fabricated.

(*b*) As a description of physical properties like rough surfaces, residual stress or random strains which may contribute to degradation.

(*c*) As a guide to how many occurrences of failure will occur, including as examples:

(i) knowing the average rate of failures (perhaps of soldering connections on an intergrated circuit) to predict how many circuits will have one failure, how many two failures, etc;

(ii) knowing the probabilities of individual contributing events, to estimate the probability of a complex rare event depending on all of them. Here one might use fault trees or event trees (§2.3);

(iii) knowing the probabilities of events of various magnitudes (e.g. earthquakes), to predict the probability of an exceptionally large event by extrapolation. This uses the statistics of extremes (Gumbel 1958) outlined later.

(*d*) To test whether a particular measurement (e.g. NDT signals from a given defect) does indeed correspond to the class of defects of concern and which have been previously characterised by NDT signals;

(*e*) To give confidence limits in verifying a hypothesis (perhaps about human factors) from limited or approximate data;

(*f*) To deduce optimal values from measurements, e.g. to characterise fracture of a particular steel.

We shall not discuss all of these uses, since several are discussed at length in the standard texts.

One cannot discuss statistics without some remarks on probability (see Parry and Winter 1980 for a fuller discussion). The common view is *empirical*, i.e. that the relative frequencies have well defined limits after many events, and that the same limits would be obtained from any randomly selected subset. There is no analysis of a single event. *Logical* probability, on the other hand, assigns (in ways not always agreed) numerical levels of belief which relate some proposed events to the body of knowledge deemed relevant. Out of this approach grew *subjective* probabilities. The Bayesian view illustrates this: one assumes a degree of belief ($P(H|E)$ in a hypothesis H given evidence E) and this $P(H|E)$, continuously revised with new data, constitutes a probability. Whereas the classical view maintains that the parameters in any situation are fixed but unknown, the Bayes approach assumes them to be random within a fixed prior distribution.

Two special cases—both quite common—need emphasis. First, suppose you find *no* failures at all. What can you say about the probability of failure? Merely that the most likely frequency of failure is zero, with some upper bound on likely rate. Secondly, suppose you find only *one* failure. What then? You now have some idea of one of the ways failure can occur—but no more. The zeroth law of statistics is this: if the removal of a single datum alters your statistical analysis appreciably, you do not have statistics and should not apply statistical arguments†.

We can see from this that statistical methods, if applicable, will depend on the size of the database: when there are no observations, we clearly have no statistics. If there are only rare observations, we follow the Bayes approach, i.e. attempt a continuous update of probabilities. As the database improves, inductive statistical methods become possible. If we reach the happy state of having many data, the approaches of data analysis can be used in full.

A.2   Aspects of standard statistical methods

A2.1   The binomial distribution

Suppose we make a measurement with two discrete possible outcomes. Perhaps we are testing a soldered connection on a semiconductor device, and it is found faulty on a fraction p of tests and acceptable in the remain-

† The apparent single observation of a magnetic monopole has provoked fascinatingly silly statistical comments. One has reason to expect a sighting every 10^5 years. A single event appears after 6 months, i.e. a lucky strike, if true. No more events are seen. This, the physicists claim, *decreases* the probability of the event being true, even though the *rate* of these events is moving steadily *towards* the expected value.

ing fraction $q \equiv (1 - p)$. We can use the binomial distribution to find the probability that there will be a proportion p of faulty devices in a small sample n out of a very large batch of N devices. Specifically, the probabilities of 0, 1, 2, ... faulty devices are given by the successive terms in the standard expansion

$$(q + p)^n = q^n + nq^{n-1}p + \frac{n(n-1)}{2} q^{n-2}p^2 + \ldots$$

where the coefficients are the standard binomial coefficients. The generalisation to several well defined outcomes gives the *multinomial* distribution, based on the expansion of $(q + p_1 + p_2 + \ldots)^n$.

A2.2   The Poisson distribution

With the binomial distribution, one took a sample of *known* size. The numbers of each possible outcome were clearly defined. This is not always so. We may know how many times lightning has struck, but not how many times it has failed to do so, for the events are discrete within the continuum of time. The Poisson distribution covers just this situation, assuming a stationary (i.e. the probabilities are constant in time) random process. If the average number of events in a given interval is M, then the probabilities of 1, 2, 3, ... events in many intervals of equal length would be

$$e^{-a}\left[1, a, \frac{a^2}{2!}, \frac{a^3}{3!} \ldots \right]$$

i.e. the successive terms in the expansion $e^{-a} e^{+a}$.

We note that if a *rare* process is defined (Janossy *et al* 1950) by the criterion

$$\underset{t \to 0}{\text{Limit}}\left(\frac{\text{(Probability of one event in time interval } 0 - t)}{\text{(Probability of at least one event in time interval } 0 - t)}\right) = 1$$

then Grigeloinis' theorem (see Thompson 1979) asserts that the superposition of a large number of rare processes gives a Poisson process (see also Uppuluri (1980) for discussion of related topics).

A2.3   The normal Gaussian distribution

It is well known by theorists that this distribution is experimentally based; experimenters know it as theoretical necessity. The reality is different: the normal (Gaussian) distribution appears as either a limiting result (the key is the central limit theorem) or a good approximation to many situations in which a large number of small contributions are combined. Thus the binomial distribution approaches the normal distribution as n becomes large (how large depends on p and q; the approach is more rapid when p and q are similar). The Poisson distribution also approaches the normal distribution for large a. The usual form of the normal distribution gives a probability $P(x)$ that some variable lies between x and $x + dx$:

$$P(x) = [2\sigma^2\pi]^{-\frac{1}{2}} \exp[-(x - \bar{x})^2/2\sigma^2]$$

defined by a mean value $\bar{x}$ and a standard deviation σ.

It is worth emphasising that normal distributions do not occur in all the cases one might have guessed. The random stresses and strains in solids due to defects are not usually Gaussian in distribution (see Stoneham (1969) for a review).

A2.4   Bayes' theorem

Here the idea, first given by Bayes posthumously in 1763, is a systematic revision of assessments of probabilities in the light of extra information. Let us write the probability of A assuming conditions B as $P(A|B)$. We can generalise this formally, writing $P(A|B)$ as the set of probabilities of the various options $A_1, \ldots, A_N$ given the situation described by B. Suppose additional information arises which we describe by α (α may represent a specific outcome like A_i in another observation, for example). Then the *product rule* for combining probabilities gives

$$P(\alpha, A|B) = P(\alpha|B)\ P(A|\alpha, B)$$
$$= P(A|B)\ P(\alpha|A, B)$$

so that the posterior probability becomes

$$P(A|\alpha, B) = [P(A|B) \cdot P(\alpha|A, B)]/P(\alpha|B).$$

If the $A_1, \ldots, A_N$ are mutually exclusive and cover all options, then

$$P(\alpha|B) = \sum_i P(A_i|B)P(\alpha|A_iB).$$

Thus if we know *prior probabilities* $P(A|B)$ from past information B, and if additional information α defines *likelihoods* $P(\alpha|AB)$ then these can be combined to give *posterior probabilities* $P(A|\alpha, B)$.

A.3   Asymptotic distributions

Many problems involve the very largest or very smallest cases of some variable which is statistically distributed. The quantity may be the lifetime of a component, the stress to cause fracture, or the dimensions of grains in a metal. In such cases the usual statistical approaches, based on the normal distribution, may be of no help. If we have results on the largest observed values from 100 tests, how can we tell anything about the tail of that part of the distribution which is less than $1/100$ of the total? This is the so-called 'statistics of extremes', reviewed in depth and clarity by Gumbel (1957).

Suppose we know the *initial probability* $F(x)$ that the variable is less than x; correspondingly $1 - F$ is the initial probability the variable exceeds x. The probability of a value between x and $x + dx$ is $f(x) = dF/dx$. The *failure* or *hazard rate*, the probability that a value, known to exceed x, actually lies between x and $x + dx$, is

$$\mu(x) \equiv f(x)/(1 - F(x)).$$

The *return period*, i.e. the frequency with which values greater than x recur, is

$$T(x) \equiv 1/(1 - F(x)).$$

All these four quantities (F, f, μ, T) are part of standard statistics.

Let us define a *characteristic largest value* u_n for n observations by $F(u_n) = 1 - (1/n)$, or $(1 - F(u_n))n = 1$. This value is the one which will be exceeded just once in n observations. Clearly it depends on n. The probability that, in any set of n observations the largest value is x (hence the probability that all other values are less than x) is $(F(x))^{n-1}$ which, in the limit of large n, tends to one of three (and only three) types.

Since the limit as n tends to ∞ of $[1 - (X/n)]^n$ is $\exp(-X)$, with a little algebra we see that in the limit $n \to \infty$,

$$(F(x))^n = [1 - (1 - F(A))]^n = \left[1 - \frac{1}{n}\left(\frac{1 - F(x)}{1 - F(u_n)}\right)\right]^n$$

$$\to \exp\left[-\left(\frac{1 - F(x)}{1 - F(u_n)}\right)\right].$$

A3.1 The first asymptote: exponential distributions

Examples include (where we have set certain arbitrary constants to unity)

$$F(x) = 1 - \exp(-x)$$

$$F(x) = 1 - A \exp(-e^x) \qquad \text{(Gompertzian or Gumbel form).}$$

The Gompertzian was first derived assuming that the 'force of mortality' in human deaths (corresponding to $\mu(x)$) rises exponentially. The same assumption appears to work for the breaking strength of (for example) rubber samples where the largest flaws determine the smallest breaking strengths. For the Gompertzian:

$$f = Ae^x \exp(-e^x)$$

$$\mu = e^x$$

$$T = \exp(e^x)/A$$

$$u_n = \ln \ln (An)$$

and the asymptotic form of $(F(x))^n$ at large n is

$$\exp\left(-\left[\frac{1 - F(x)}{1 - F(u_n)}\right]\right) = \exp\left(-\frac{\exp(-e^x)}{\exp(-u_n)}\right).$$

The probability of survival is determined by $(1 - F(x))$. This is shown for the load/survival of rubber samples (Gumbel 1958 p. 250) where x is now the load; Doremus (1983) shows that this distribution also describes the fracture strengths of abraded glass fibres.

A3.2   The second asympote: Cauchy, Pareto and Weibull distributions

Useful cases include forms like

$$F(x) = 1 - (x - \varepsilon)^{-k}$$

where we note a power law, with a chosen minimum value for $x(x \geqslant \varepsilon)$ and $k \geqslant 1$, $\varepsilon \geqslant 0$.

$$f(x) = k/(x - \varepsilon)^{k+1}$$

$$\mu(x) = k/(x - \varepsilon)$$

$$T(x) = (x - \varepsilon)^k$$

$$(u_n - \varepsilon)^k = 1/n$$

with the asymptotic form

$$\left[\exp - \left(\frac{x - \varepsilon}{u_n - \varepsilon}\right)^{-k}\right].$$

Here we have a three-parameter distribution: a minimum lifetime ε, a scale parameter k affecting the distribution shape, and a characteristic number u_n. These distributions have been used widely to describe fatigue (Gumbel 1958, §7.3.5), fracture (Davidge 1979) and flaw size distributions (Ichikawa 1984, Ishii 1973).

The Weibull model is sufficiently important as a rationalisation of this type of distribution that we outline it here. Suppose we have a system under stress in which there are N links in series, i.e. such that failure of one link causes the whole to fail. Suppose further that we know $F_1(\sigma)$, the probability that one link will fail first under stress σ. For the N links, the weakest one is critical, so (since $1 - F_1(\sigma)$ is the probability of survival of one link) we have

$$F_n(\sigma) = 1 - (1 - F_1(\sigma))^N \underset{(N \to \infty)}{\to} 1 - \exp(-NF_1(\sigma)).$$

Suppose we write $NF_1(\sigma)$ as a product of a volume (since it is proportional to the number of links) and a function of σ, which we assume to be of the form $[(\sigma - \sigma_u)/\sigma_0]^m$ we gain the well known Weibull form which, with changes of notation ($\sigma \to x$, $\varepsilon \to \sigma_u$, $u_n \to \sigma_0 + \sigma_u$, $k \to -m$) this gives the second asymptotic distribution. As m gets larger, the more abrupt becomes the survival probability at $\sigma = \sigma_u + \sigma_0$, becoming a step function as $m \to \infty$. Typical values for m are

$\sim$ 2–3	glass
$\sim$ 10	Si_3N_4 at high temperatures $T > 1300\,^{\circ}C$
$\sim$ 12	graphite
$\sim$ 13	alumina
$\sim$ 40	cast iron
$\sim$ 70	Si_3N_4 at low temperatures.

A3.3   Outliers in statistics

Our discussion of extreme values discusses what one could expect from observations of normal values. Outliers are apparently abnormal. It is a related issue as to whether they should be accepted as orthodox members of a distribution (like being dealt a hand consisting solely of spades from a shuffled pack) or whether they are anomalous, to be rejected for reasons beyond mere prejudice. We merely cite Earnett and Lewis (1978) for a comprehensive discussion.

A.4   Mathematical description of unreliability

Consider a non-destructive examination in which some parameter, like the amplitude of a signal on an oscilliscope screen, is compared with that from a reference reflector. Let us call the amplitude from the reference reflector A_c and that from the 'defect' A_s then, according to the inspection procedure rules, we report the defect if $A_s \geqslant A_c$. However, in practice both A_s and A_c will be different each time they are measured, even for the same defect and the same calibration reflector. This variability is common to any physical measurement and is represented by using a distribution about some mean value. A useful (but by no means the only useful) distribution is the normal or Gaussian distribution. In this, the probability of obtaining a value lying between x and $x + dx$ for some parameter x is given by $P(x)dx$

where

$$P(x) = \frac{1}{\sigma\sqrt{(2\pi)}} \exp\left[-\frac{1}{2}\left(\frac{x-\mu}{\sigma}\right)^2\right]$$

where μ is the mean and σ the standard deviation of this distribution. Suppose that both the actual defect signal and that from the calibration reflector are given by this type of distribution; this should often be a satisfactory hypothesis and is used for illustrative purposes here. We assume also that the two distributions have the same standard deviation σ; this simplifies the analysis as well as being a justifiable approximation in many cases. The means from the defect and calibration experiments are μ_s and μ_c respectively. A defect goes unreported if $A_s < A_c$. In figure 5.6 we show means of the hatched area the region in which, although the defect is actually bigger than the calibration and ought to be reported, it gives a smaller signal. The probability of incorrect reporting, in this case is $B(x)$ say, where

$$B(x) = \int_{-\infty}^{d} \frac{1}{\sigma\sqrt{(2\pi)}} \exp\left[-\frac{1}{2}\left(\frac{x_s-\mu_s}{\sigma}\right)^2\right] dx_s$$

which can be written as

$$B(x) = 1 - 0.5\ \mathrm{erfc}(d)$$

where erfc is the complementary error function. The value of d is obtained from

$$d = (\mu_c + \mu_s)/(2\sigma)$$

and a plot of the function $B(x)$ is given in figure 5.7, where we see that it has a characteristic S shape. The method we have indicated above is related to 'signal detection theory' which is used to estimate the failure to detect and correctly classify rate and, conversely, the false call rate. In terms of the notation commonly employed in that field, our parameter d is the β of signal detection theory, where β stands for bias. Thus, in our case we have set the boundary of false calls to detection failures midway between the two means implying zero bias towards either false calls or detection failures. The distance between the means, in standard error units, is usually denoted by the symbol d' in the literature on signal detection theory. This forms the basis of receiver operating characteristics (ROC) and their use in estimating the reliability of detection equipment.

A4.1   Defect sizing errors based on measurement of defect extremities

Another problem, related to the above, is that of measuring a defect size from two measurements of position of the defect ends. This is common practice in most conventional ultrasonic inspection techniques and also in the time-of-flight ultrasonic technique, for example. What is usually important in these inspections is the ability to correctly reject all defects above some specified size. Let us assume, for the present, that the distribution of measured position about a mean at each end of the defect is normally distributed and that the means of the distributions accurately represent the actual positions of the ends of the defect. Suppose we know that the distributions at each end are Gaussian with means μ_1 and μ_2 respectively, and that the standard deviation of both distributions is equal to σ. This could represent the measurement of defect through-wall size by time-of-flight, for example, and is illustrated in figure 5.8. For the present let us take these means μ_1 and μ_2 to be accurate representations of the extremities of the actual defect. The probability of obtaining a position lying between x and $x + dx$ for the first extremity of the defect is then $P_1(x) \, dx$ where

$$P_1(x) = \frac{1}{\sqrt{\pi}} \exp[-(x - \mu_1)^2].$$

We have measured the means and position in terms of normalised quantities. That is $x \to x/\sigma\sqrt{2}$ and $\mu \to \mu/\sigma\sqrt{2}$. Now suppose that we are interested in the probability that a measured value will exceed some specified value x_c. This specified value could be that specified by a critical size above which the component might be unreliable as determined by fracture mechanics, for example. Again we are talking in terms of a normalised value $x_c \to x_c/\sigma\sqrt{2}$. We take any value in the region x to $x + dx$ and calculate the conditional probability $P(s > x_c \mid x)$ that the measured size s exceeds the critical size x_c. This is given by

$$P(s \geqslant x_c \mid x) = P_1(x) \int_{x+x_c}^{\infty} P_2(y) \, dy + P_1(x) \int_{-\infty}^{x-x_c} P_2(y) \, dy.$$

Since any value of x was possible with its associated probability we must sum this last expression over all valid values to give

$$P(s \geqslant x_c) = \int_{-\infty}^{\infty} P_1(x) \left(\int_{x+x_c}^{\infty} P_2(y) \, dy + \int_{-\infty}^{x-x_c} P_2(y) \, dy \right) dx.$$

Looking at the inner integrals we observe that

$$\int_{x+x_c}^{\infty} P_2(y) \, dy = \tfrac{1}{2} \, \text{erfc}(x + x_c - \mu_2)$$

and, similarly,

$$\int_{-\infty}^{x-x_c} P_2(y) \, dy = \tfrac{1}{2} \, \text{erfc}(-x + x_c + \mu_2).$$

Hence our probability of measuring the defect to be larger than x_c becomes

$$P(s \geqslant x_c) = \frac{1}{2\sqrt{\pi}} \int_{-\infty}^{\infty} \exp - (x - \mu_1)^2$$
$$\times \, [\text{erfc}(x + x_c - \mu_2) + \text{erfc}(-x + x_c + \mu_2)] \, dx.$$

Note, in passing, that if $x_c = 0$ and $\mu_2 = 0$ this is just

$$\frac{1}{2\sqrt{\pi}} \int_{-\infty}^{\infty} \exp - (x - \mu_1)^2 \; [\mathrm{erfc}(x) + \mathrm{erfc}(-x)] \; \mathrm{d}x$$

and

$$\mathrm{erfc}(-x) = 2 - \mathrm{erfc}(x)$$

so this becomes

$$\frac{1}{\sqrt{\pi}} \int_{-\infty}^{\infty} \exp[-(x - \mu_1)^2] \; \mathrm{d}x = 1$$

as expected, since we are certain to measure a modulus of the defect size to be bigger than, or equal to, zero. Plots of $1 - P(s \geqslant x_c)$ are given in figure 5.9 for $x_c/\sigma\sqrt{2} = 0$, 0.75, 2, 4, 6 and for $0 \leqslant |(\mu_2 - \mu_1)/\sigma\sqrt{2}| \leqslant 9$.

A4.2 Defect under or over sizing

Another interpretation can be placed on these results, where previously we took the two means μ_1 and μ_2 to be accurate representations of the actual defect extremities and the parameter x_c to be specified as a critical size, we can also take x_c to represent the actual defect size and μ_1, μ_2 to be experimental values. Then we can associate $P(s \geqslant x_c)$ as the probability of measuring a defect which is bigger than, or equal to, the actual defect size. For example, from figure 5.9 we see that for a real defect size $x_c = 4\sigma\sqrt{2}$ and if the measurements give $|\mu_2 - \mu_1| = 3\sigma\sqrt{2}$ then the likelihood that we measure a defect bigger than the real defect is only about 0.16. To estimate what the chance of measuring $s \equiv x_c$ we would need to proceed as follows

$$P(s \equiv x_c) = \int_{-\infty}^{\infty} [P_1(x)P_2(x + x_c) + P_1(x)P_2(x - x_c)] \; \mathrm{d}x$$

using the useful relationship (Gradshteyn and Ryzhik 1980, §3.323), provided $p > 0$, that

$$\int_{-\infty}^{\infty} \exp(-p^2x^2 \pm qx) \; \mathrm{d}x = \exp(q^2/4p^2) \frac{\sqrt{\pi}}{p}$$

which gives

$$P(s \equiv x_c) = \frac{1}{\sqrt{2}} \exp[-\tfrac{1}{2}(\mu_1 - \mu_2 + x_c)^2] + \frac{1}{\sqrt{2}} \exp[-\tfrac{1}{2}(\mu_1 - \mu_2 - x_c)^2].$$

In the above discussions, the significance of σ is that it represents the precision with which the measurements can be repeated, a small σ indicating a high degree of repeatability of that inspection. The distributions need not be symmetrical but a normal distribution is representative for the sort of discussion of errors likely in the measurement of a signal amplitude or the position of an extremity of a defect.

References

Aaltio M and Kauppinen K P 1982 Reliability and defect sizing *Proc. Conf. Periodic Inspection of Pressurized Components* (London: IME) pp 283–90

Adamenko A, Vallevich M I and Demidko V G 1979 Evaluation of the detectability of real flaws by the results of detection of reference flaws in radiographic inspection of welded joints *Sov. J. NDT* **15** 601

Aldridge E E 1969a Ultrasonic holography *Engineering* **208** 71–4

—— 1969b Ultrasonic holography *Phys. Bull.* **20** 10–12

—— 1970 Ultrasonic holography in *Research Techniques in Nondestructive Testing* ed R S Sharpe (London: Academic) pp 133–54

Aldridge E E and Clare A B 1980 Ultrasonic holography for the inspection of thick plate structures *CEC Report* EUR 6865

Aldridge E E, Clare A B, Lloyd G A S, Shepherd D A and Wright J E 1971 A preliminary investigation of the use of ultrasonic holography in ophthalmology *Br. J. Radiol.* **44** 126

Aldridge E E, Clare A B and Shepherd D A 1970 Ultrasonic holography in nondestructive testing *Acoustical Holography* vol. 3 ed A F Metherell (New York: Plenum) pp 129–45

—— 1974 Scanned ultrasonic holography for ophthalmic diagnosis *Ultrasonics* **12** 155–60

Aldridge E E and Clement M J 1982 Ultrasonic holography *Ultrasonic Testing – Non-conventional Testing Techniques* ed J Szilard (New York: Wiley) p 103

American Society for Metals 1970 *Metals Handbook* 8th edn part 5B

Ando Y, Tsubota M and Takama S 1982 Reliability assessment activities of Japan in relation to in-service inspection in *Periodic Inspection of Pressurized Components* (London: Institute of Mechanical Engineers) pp 295–303

Apostolakis G 1978 Philosophical aspects of rare event probabilities *Nucl. Safety* **19** 305

Arakawa T 1983 *Mater. Eval.* **41** 714–19

ASME (American Society of Mechanical Engineers) 1974, 1977, 1983 *Pressure Vessel and Boiler Code* and semi-annual addenda (the main difference between the 1983 edition and the earlier editions is the introduction of SI units)

ASNT Publications Committee 1980 *Mater. Eval.* **38** 25–31

Atkinson P and Berry M V 1974 Random noise in ultrasonic echoes diffracted by blood *J. Phys. A: Math. Gen.* **7** 1293

Bailey R W 1935 *Proc. Inst. Mech. Eng* **131** 131

Baker C H 1958 Attention to visual displays during a vigilance task. 1 Biassing attention *Br. J. Psychol.* **49** 279–88

Baker C H and Boyes G E 1959 Increasing probability of target detection with a mirror-image display *J. Appl. Psychol.* **43** 195–8

Baker C H, Ware J R and Sipowicz S 1962 Sustained vigilance: 1. Signal detection during twenty-four hour continuous watch *Psychol. Rec.* **12** 245–50

Barbian O A, Engl G, Grohs B, Rathgeb W and Wüstenberg H 1984 A second view of the German results in the defect detection trials, UKAEA *Br. J. NDT* **26** 92–6

Barbian O A, Grohs B, Kappes W and Hullin Ch 1983 Inspection of thick-walled components by ultrasonics and the evaluation of the data by the ALOK technique in *New Procedures in NDT* ed P Holler (Berlin: Springer) p 133

Barbian O A, Grohs B and Neumann R 1982 ISI by ultrasonics in nuclear power plants using the ALOK technique *Periodic Inspection of Pressurized Components* (London: IME) pp 197–208

Barenblatt G I 1962 *Adv. Appl. Mech.* **7** 55

Barley R E and Proschan F 1965 *Mathematical Theory of Reliability* (New York: Wiley)

Barnett V and Lewis T 1978 *Outliers in Statistical Data* (New York: Wiley)

Barton C J E 1982 Some legal aspects of fitness for purpose in *Fitness for Purpose Validation of Welded Constructions* (Abington, Cambridge: The Welding Institute) pp 3.1–3.4

Bayes T 1763 *Phil. Trans. R. Soc.* **53** 370

Becker F L 1980 Ultrasonic inspection reliability for primary piping systems paper presented at *CSNI Specialist Meeting on Reliability of Ultrasonics Inspection of Austenitic Materials and Components, Brussels 29–30 May, 1980*

Becker F L, Doctor S R, Heasler P G, Morris C J, Pitman S G, Selby G P and Simonen F A 1981 Integration of NDE reliability and fracture mechanics *NUREG Report* CR-1696 PNL 3469 vol. 1

Becker R and Betzhold K 1983 Defect classification by multifrequency eddy currents in *New Procedures in NDT* ed P Holler (Berlin: Springer) p 497

Berens A P and Hovey P W 1982 Characterization of NDE reliability in *Review of Progress in Quantitative NDE* vol. 1 ed D O Thompson and D E Chimenti (New York: Plenum)

—— 1983 Statistical methods for estimating crack detection probabilities *ASTM STP 798* ed J M Bloom and J C Ekvall (Philadelphia: American Society for Testing and Materials) pp 79–94

Berger H 1965 *Neutron radiography. Methods, capabilities and applications* (Amsterdam: Elsevier)

—— (ed) 1976 *Practical Applications of Neutron Radiology and Gaging. Proc. Symp. on Practical Applications of Neutron Radiography and Gaging, Gaithersburg, USA, 10–11 February 1975* ASTM Special Technical Publication 586 (Philadelphia: ASTM)

—— (ed) 1977 *Nondestructive Testing Standards: a review. Proc. Symp. on Nondestructive Testing Standards, Gaithersburg, 19–21 May 1976* ASTM Special Technical Publication 624

Bertram W J 1983 in *VLSI Technology* ed S M Sze (New York: McGraw-Hill) p 599

Betz C E 1963 *Principles of Penetrants* (Chicago: Magnaflux Corp.)

Billington R and Allan R N 1983 *Reliability Evaluation of Engineering Systems: Concepts and Techniques* (London: Pitman)

Biloni H 1983 *Solidification of Physical Metallurgy* ed R W Cahn and P Hadsen Ch. 9 (Amsterdam: Elsevier)

Blistanova T A, Mikhailov S K, Rozina M V, Rybin Yu I and Yablonik L M 1977 *Sov. J NDT* **13** 657–61 (translated from *Defektoskopyia* **6** 62–8 (1977))

Bogie K D 1982 *Paper Presented at the British Institute of Non-destructive Testing Annual Conference NDT '82, York 20–22 September, 1982*

Bogie K D and Beyers C J E 1982 *Paper Presented at the British Institute of Non-destructive Testing Conference NDT '82, York 20–22 September, 1982* (published in *Br. J. NDT* **26** 218–22 (1984))

Boness K D 1982 A pulsed tuned eddy current system for NDT *UKAEA Report* AERE – M 3245

Bortz S 1979 *Army Mater. Conf. Ser.* **6** 445

Bott C H 1983 Non-destructive testing standards and codes in the aerospace industry *NDT Aust.* **20**(4) 12–16

Bowker K J, Coffey J M, Hanstock D J, Owen R C and Wrigley J M 1983 CEGB inspection of plates 1 and 2 in UKAEA defect detection trials *Br. J. NDT* **25** 249–55

Bowman K O and Leach E 1983 *Oak Ridge National Laboratory Report* ORNL/CSD-115

British Standards 1983 *An Index and Guide to Publications* (This gives the following reference numbers for relevant topics in NDT: ultrasonic flaw detection – calibration blocks 2704, – performance tests 4331, testing of pipes 3889, testing of plate material 5996, testing of steel castings and forgings 4080, 4124, and for radiographic inspection we have – testing of butt joint 2600, inspection of circumferential welds 2910, steel castings 4080)

British Standards Institution (BSI) 1983 Quality assurance. First revision *BSI Handbook* 22 (lists all the relevant control documents for the NDT techniques)

Buehler M G 1983 in *VLSI Technology* ed S M Sze (New York: McGraw-Hill) p 548

Buntin W D 1972 Concept and conduct of proof test of F-111 production aircraft *Aeronaut. J.* **76** 587–98

Burch S F 1980 Application of digital techniques to the restoration of radiographic images *UKAEA Report* AERE-R 9833 (London: HMSO)

—— 1984 Digital image processing in nondestructive testing in *Research Techniques in Nondestructive Testing* vol.7 ed R S Sharpe (London: Academic) pp 1–35

—— 1985 An amplitude correlation and differencing method for the monitoring of flaws in repeat ultrasonic inspections *Ultrasonics* **23** 246–85

Burgess N T (ed) 1983 *Quality Assurance of Welded Construction* (New York: Applied Science)

Bush S H 1979 *Battelle Pacific North-West Report* PNL-SA-8090

—— 1981 *Non-destructive Evaluation: Proc. AIME Metallurgical Society Symp. at Pittsburgh 5–9 October, 1980* ed O Buck and S M Wolf (Warrendale, PA: AIME) pp 65–79

Calderola L 1980 *Nucl. Eng. Des.* **58** 127

Cameron A G B, Jessop T J, Smith P H and Gibson R 1981 Extended studies on ultrasonic detection and measurement of fatigue cracks in mild steel *Welding Institute Report* 3618/1/81

Cameron R F 1984 Theoretical calculations of pressure vessel failure frequencies for Sizewell B transients *UKAEA Report* AERE-R 11196 (London: HMSO)

Cameron R F, Johnson G O and Lidiard A B 1986 The reliability of pressurized water reactor vessels in *Probabilistic Fracture Mechanics* ed J W Provan (Dordrecht: Martinius Nijhoff)

Cameron R F and Temple J A G 1985 Ultrasonic inspection for long defects in thick steel components *Int. J. Pres. Ves. Piping* **18** 255–76

—— 1986 A method of quantifying the reliability required of non-destructive inspection of PWR pressure vessels *Nucl. Eng. Des.* **91** 57–68

Carstensen E L and Schwan H P 1959 Acoustic properties of haemoglobin solutions *J. Acoust. Soc. Am.* **31** 305

Cattiaux A M, Morisseau G, Pincemille P and Saglio R 1983 An in-service method: the use of focussed probes for detection and sizing in DDT plates 1 and 2 *Br. J. NDT* **25** 258

Caussin P, Cermak J and Verspeelt D 1980 *Proc. Specialists' Meeting on Reliability of Ultrasonic Inspection of Austenitic Materials, Brussels, May 29–30, 1980* (London: Applied Science)

Caussin P and Verspeelt D 1981 *Proc. 4th Int. Conf. on Non-destructive Testing in the Nuclear Industry, Lindau, West Germany, May 25–27, 1981* (Berlin: DGfZP)

CEC/OECD 1985 Commission of the European Communities and Organization for Economic Co-operation joint reports by the Nuclear Energy Agency Committee on the Safety of Nuclear Installations *Report CSNI 106* A summary of the PISC-II project, *CSNI 107* The round-robin test of the PISC-II programme: plates and ultrasonic procedures used, *CSNI 108* Destructive examination of the PISC-II RRT Plates, *CSNI 109* Analysis scheme of the PISC-II trials results, *CSNI 110* First evaluation of the PISC-II trials results

—— 1986 Commission of the European Communities and Organisation for Economic Co-operation joint reports by the Nuclear Energy Agency Committee on the Safety of Nuclear Installations *Report CSNI 117* A summary of the PISC-II project, *CSNI 118* The round-robin test of PISC-II programme: plates and ultrasonic procedures used, *CSNI 119* Destructive examination of the PISC-II RRT plates, *CSNI 120* Analysis scheme of the PISC-II trials results, *CSNI 121* Evaluation of the PISC-II trials results

Central Statistical Office 1983 *Annual Abstract of Statistics* (London: HMSO)

Chaloner W H 1959 *VULCAN: The history of one hundred years of engineering and insurance 1859–1959* (Manchester: Vulcan Boiler and General Insurance Co Ltd)

Charlesworth J P and Hawker B M 1984 Inspection of the near-surface defect plate (DDT3) by the ultrasonic time-of-flight technique *Br. J. NDT* **26** 106–11

Charlesworth J P and Temple J A G 1981 Creeping waves in ultrasonic NDT in *Ultrasonics International '81* (Sutton, Surrey: IPC Science and Technology Press) pp 390–5

—— 1982 Ultrasonic time-of-flight inspection through anisotropic cladding in *Periodic Inspection for Pressurized Components* (London: IME) pp 117–24

Chegorinski V A 1971 *Sov. J. NDT* **7** 357–9 (translated from *Defektoskopyia* **3** 136–8 (1971))

Chesney D N and Chesney M O 1981 *Radiographic Imaging* 4th edn (Oxford: Blackwell Scientific)

Chin-Quan H R and Scott I G 1977 Operator performance and reliability in NDI in *Research Techniques in NDT* vol.3 ed R S Sharpe (New York: Academic) pp 323–54

Chockie L J 1978 in *Non-destructive Examination in Relation to Structural Integrity* ed R W Nichols (New York: Applied Science) p 116

Coffey J M 1978 *Proc. Conf. on Tolerance of Flaws in Pressurized Components, 16–18 May London* (London: IME) pp 57–63

—— 1979 Introductory review *Br. Inst. NDT Symp. on Improving the Reliability of Ultrasonic Inspection, Northampton, 1979* (Northampton: British Institute of NDT)

—— 1982 The reliability of ultrasonic inspection for thick section welds: some views and model calculations paper C159/82 in *Periodic Inspection of Pressurized Components* (London: IME) pp 273–82

Coffey J M, Chapman R K and Hanstock D J 1982 The ultrasonic detectability of a postulated 'worst case' flaw in a PWR pressure vessel' *CEGB Report* NW/SSD/82/0045/R April 1982 (London: CEGB)

Collier J G 1983 Reliability problems of heat transfer equipment *Atom* **322** 172

Colquhoun W P and Edwards R S 1970 Practice effects on a visual vigilance task with and without search *Human Factors* **12** 537–45

Comassar D M 1978 *State of the Art of Nondestructive Inspection of Aircraft Engines* AGARD-LS-103

Conrad R 1951 Speed and load stress in sensory-motor skill *Br. J. Indust. Med.* **8** 1–7

Cook D 1972 Crack depth measurement with surface waves *Br. Acoust. Soc. Proc.* **1** (Spring Meeting 5–7 April, 1972. University of Loughborough. Ultrasonics in Industry Session. paper No. 72U19)

Cooper M J and Cheeseman M 1977 Reliability and architecture of plant safety systems *UKAEA Report* AERE – R 8959

Cottrell A H 1964 *The Mechanical Properties of Matter* (New York: Wiley) p 329, 346

Crutzen S, Bürgers W, Violin F, Di Piazza L, Cowburn K J and Sargent T 1983 *Br. J. NDT* **25** 193–4

CSNI 1978 *Statistics and Decision Theories for Rare Events* SINDOC(78) 85A

Curran D R, Seaman L and Shockey D A 1977 *Physics Today* (January) 46

Currie T 1983 The Need For Validation of Techniques and Procedures presented at the *OECD/NEA Specialist Meeting on Ultrasonic Defect Detection and Sizing, Ispra, May 3–6, 1983* (Paris: OECD Nuclear Energy Agency) pp 356–66

Curtis G J and Hawker B M 1983 Automated time-of-flight studies of the defect detection trial plates 1 and 2 *Br. J. NDT* **25** 240–8

Danchak M M 1984 Human factors of CRT displays for nuclear power plant control *Adv. Nucl. Sci. Tech.* **16** 75

Davidge R W 1979 *Mechanical Behaviour of Ceramics* (Cambridge: Cambridge University Press)

Davies D R and Hockey G R J 1966 The effects of noise and doubling the signal frequency on individual differences in visual vigilance performance *Br. J. Psychol.* **57** 381–9

Deane J A 1979 The influence of component surface finish on ultrasonic test performance *Br. Inst. NDT Symp. on Improving the Reliability of Ultrasonic Inspection, Northampton 1979* (Northampton: British Institute of NDT)

Dieulsaint E and Royer D 1980 *Elastic Waves in Solids* (Chichester: Wiley)

Doctor S R, Becker F L, Heasler P G and Selby G P 1983 Effectiveness and reliability of inservice inspection, a round robin test presented at the *OECD/NEA Specialist Meeting on Ultrasonic Defect Detection and Sizing, Ispra, May 3–6 1983* (Paris: OECD Nuclear Energy Agency)

Doctor S R, Selby G P, Heasler P G and Becker F L 1982 Effectiveness and reliability of inservice inspection, a round robin test *NDE in the Nuclear Industry. Proc. 5th Int. Conf., San Diego, May 1982* (Ohio: American Society for Metals)

Dolby R E 1981 Data correlation of crack opening displacement and Charpy V test data *Welding Institute Report*

Doremus R H 1983 *J. Appl. Phys.* **54** 153

Dover W D, Charlesworth P D W, Taylor K A, Collins R and Michael D H 1979 The use of ac field measurements to determine the shape and size of cracks in metal in *Proc. Symp. on Eddy Current Characterisation of Materials and Structures, Gaithersburg*

Druce S G and Edwards B C 1980 *AERE Report* R9652

Druce S G and Hudson J A 1982 Defects arising from welding and cladding PWR pressure vessel steels: a review *AERE Report* R10418

Dugdale D S 1960 *J. Mech. Phys. Sol.* **8** 100

Easterling K 1983 *Introduction to the Physical Metallurgy of Welding* (London: Butterworths)

Edwards G T and Watson I A 1979 *UKAEA Safety and Reliability Directorate Report* SRD R 146

Elkin E H 1958 Effect of scale shape, exposure time and display complexity on scale reading efficiency *USAF, WADC, Technical Report* 58-472

Ely R V 1980 *Microfocal Radiography* (London: Academic)

Embrey D E 1976 Human reliability in complex systems: an overview *National Centre of Systems Reliability Report* NCSR R 10

—— 1981 Success—likelihood index *3rd National Reliability Conference, Birmingham* (London: HMSO)

Erhard A, Wustenberg H, Engle G and Kutzner J 1980 *NDE in the nuclear industry. Proc. 3rd Int. Conf., Salt Lake City, February 1980* (Ohio: American Society for Metals) pp 255–68

Erhard A, Wustenberg H and Kutzner J 1979 *Br. J. NDT* **21** 39–43

Fairburn W 1864 *Phil. Trans. R. Soc.* **154** 311

Farrer J C M 1982 *Fitness for Purpose Validation of Welded Constructions, Proc. Int. Conf. on Fitness for Purpose Validation of Constructions, London 17–19 November 1981* vol. 1 (Abington, Cambridge: The Welding Institute) paper P 14.1

Fischoff B, Stovic P, Lichtenstein A, Read S and Combs B 1978 *Policy Sci.* **9** 127

Fordham P 1968 *Nondestructive Testing Techniques* (London: Business Publications)

Førli O 1979 Reliability of ultrasonic and radiographic weld testing *Proc. 9th World Conf. on NDT, Melbourne, 19–23 November 1979* paper 3A. 7

Forsten J and Aaltio M 1982 On the detection of weld defects by X-ray and ultrasonic examination *Br. J. NDT* **24** 33–6

Forster F 1980 *Proc. Symp. New Methods of NDE of Materials and their Application Especially in Nuclear Engineering* (Berlin: DGfZP) p 129

Frederick J R, Dixon M, Vanden-Broek C, Papworth D, Elzinga M, Hamano N and Ganapathy K 1978 Improved ultrasonic nondestructive testing of pressure vessels (synthetic aperture focussing technique SAFT) (October 1, 1977 – September 30, 1978) *Report* NUREG/CR 0581

Fuchs E 1961 Quality in welding *Br. Weld. J.* **8** 4

Fung K and Jardine A K S 1982 *Microelectr. Reliab.* **22** 681

Gange J J, Geen R G and Haskins S 1979 Autonomic differences between extroverts and introverts during vigilance *Psychophysiology* **16** 392–7

Gardner W E and Hudson J A 1983 Ultrasonic inspection of thick section pressure vessel steel by the time-of-flight diffraction method in *Quantitative NDE in the Nuclear Industry* ed R B Clough (Ohio: American Society for Metals) pp 250–7

Golovkin A M 1973 *Sov. J. NDT* **9** 302–6

Gordon J E 1978 *Structures, or Why Things Don't Fall Down* (Harmondsworth: Penguin) p 104

Gradshteyn I S and Ryzhik I M 1980 *Table of Integrals Series and Products* (New York: Academic)

Graff K F 1975 *Wave Motion in Elastic Solids* (Oxford: Clarendon)

Grant I M and Rogerson J H 1982 The importance of contractural requirements in determining quality costs in the fabrication industry in *Fitness for Purpose Validation of Welded Constructions Proc. Int. Conf., London 17–19 November 1981* (Abington, Cambridge: The Welding Institute) pp 9.1–9.7

Green A E 1983 *Safety Systems Reliability* (New York: Wiley)

Green A E and Bourne A J 1972 *Reliability Technology* (New York: Interscience)

Grohs B, Barbian O A and Kappes W 1983 ALOK – principles and results obtained in DDT in *Defect Detection and Sizing. Proc. OECD/NEA Specialist Meeting on Ultrasonic Defect Detection and Sizing, Ispra, 3–6 May 1983* (Paris: OECD Nuclear Energy Agency) pp 689–719

Groocock J M 1980 Quality cost control in ITT Europe *Quality Assurance* **6**(3)

Gruber G J 1983 Sizing of near-surface fatigue cracks in cladded pressure vessels by the multiple beam satellite pulse technique in *New Procedures in NDT* ed P Holler (Berlin: Springer)

Gumbel E J 1958 *Statistics of Extremes* (New York: Columbia University Press)

Gurney T R 1979 *Fatigue of Welded Structures* (Cambridge: Cambridge University Press)

—— 1983 *Proc. R. Soc.* A **386** 393

Guy A G 1971 *Introduction to Materials Science* (New York: McGraw-Hill)

Guy A G and Hren J J 1974 *Elements of Physical Metallurgy* (Reading, MA: Addison Wesley)

Hagemaier D J, Klima S and Rummel W D 1978 Reliability measurement modelling *Proc. Government/Industry Workshop on Reliability of Non-Destructive Inspections* US Government report AD-A068223, 223–242

Hagen E W 1982 *Nuclear Safety* **23** 665–8

Hahn G J and Shapiro S S 1967 *Statistical Methods in Engineering* (New York: Wiley)

Haines N F 1977 The reliability of ultrasonic inspection in *Reliability Problems of Reactor Pressure Components. Proc. Symp. on Application of Reliability Technology to Nuclear Power Plants, Vienna, 10–13 Oct. 1977* vol. 2 (Vienna: IAEA) pp 341–58 (see also *CEGB Report* RD/B/N4123)

—— 1980 The theory of sound transmission and reflection at contacting surfaces *CEGB Report* RD/B/N4744

Haines N F, Langston D B, Green A J and Wilson R 1982 An assessment of the reliability of ultrasonic inspection methods in *Periodic Inspection of Pressurized Components* (London: Institute of Mechanical Engineers) pp 239–255

Hall S F 1984 MULTIPLET – a program for calculating large event trees *UKAEA Report* SRD-R-300

Halmshaw R 1966 *Physics of Industrial Radiography* (London: Heywood)

Hansen B 1983 Statistical aspects of the evaluation of non-destructive testing reliability paper presented at the *OECD/NEA Specialist Meeting on Ultrasonic Defect Detection and Sizing, Ispra, May 3–6 1983* (Paris: OECD Nuclear Energy Agency)

Harker A H 1984 Numerical modelling of the scattering of elastic waves in plates *AERE Report* AERE TP.1059

Harris D O 1977 A means of assessing the effects of NDE on the reliability of cyclically loaded structures *Mater. Eval.* **35** 57–65

—— 1982 Applications of a fracture mechanics model of structural reliability to the effects of seismic events on reactor piping *Prog. Nucl. Energy* **10** 125–59

Harrop L P and Lidiard A B 1979 Enhancement of reliability of reactor pressure vessels by in-service inspection *UKAEA Report* TP 777

Hayes W and Stoneham A M 1985 *Defects and Defect Processes in Non-Metallic Solids* (New York: Wiley)

Haywood C R 1952 *An Outline of Metallurgical Practice* (New York: Van Nostrand)

Herr J C and Marsh G L 1978 NDT reliability and human factors *Mater. Eval.* **36** 41–6

Hossfield F, Mika K and Plesser-Walk E 1975 Bayes analysis *Julich Report* JUL 1249 (75)

Hudson J A 1980 *The Excitation and Propagation of Elastic Waves* (Cambridge: Cambridge University Press)

Hurst D P and Temple J A G 1982 Calculation of the velocity of creeping waves and their application to nondestructive testing *Int. J. Pres. Ves. Piping* **10** 451–64

Hutton P H and Schwenk E P 1978 Interpretation of acoustic emission data to indicate flaw significance in *Proc. Conf. on NDE in the Nuclear Industry* (Ohio: American Society for Metals) pp 261–9

IAEA 1983 data published in *Atomwirtshaft* November. Note that 'in operation' includes all reactors which have at some time operated except those permanently shut down and includes five reactors in OECD countries which have been subject to longer term shutdowns pending completion of remedial work or licensing procedures

Ichikawa M 1984 *Reliab. Eng.* **9** 221

International Institute of Welding *Handbook on the Ultrasonic Examination of Austenitic Welds*

Ishii Y 1973 *Non-Destructive Testing Technology* (Tokyo: Sanpo)

Jamison T D and Dau G 1981 *EPRI Report* NP-2088

Jamison T D and McDearman W R 1981 Studies on section XI ultrasonic repeatability *EPRI Report* NP-1858

Janossy L, Renyi A and Aczel A 1950 *Acta. Math. Sci. Hung.* **2** 83

Jenkins H M 1958 The effect of signal rate on performance in visual monitoring *Am. J. Psychol.* **71** 647–61

Jessop T J 1979a Size measurement and characterization of weld defects by ultrasonic testing, Parts 1 and 2 *Welding Institute Report* 3527

—— 1979b Studies on ultrasonic detection and measurement of fatigue cracks in mild steel *Welding Institute Report* 3563/3

Johnson C A 1978 in *Fracture Mechanics of Ceramics* vol. 5 ed R C Bradt, A G Evans, D P H Hasselman and F C Large (New York: Plenum) p 365

Johnson D P 1976a *Mater. Eval.* **34** 121–6, 136

—— 1976b *EPRI Report* NP-315

Johnson D P Toomay T L and Davis C S 1979 Estimation of the defect detection probability for ultrasonic inspection on thick section steel weldments *EPRI Report* NP-911

Johnson H H and Paris P C 1968 *Eng. Fract. Mech.* **1** 3

Jones P M S 1984 *Atom* **377** 29

Kearns W H (ed) 1978 *Welding Handbook* (especially vols 1, 2, 3), (Miami: American Welding Society; London: Macmillan)

Kendall K 1983 *Am. Inst. Phys. Conf. Proc.* **107** 78

Kendall Sir M and Stuart A 1977 *The Advanced Theory of Statistics* vol. 1 Distribution theory, 4th edn (London: Griffin)

—— 1979 *The Advanced Theory of Statistics* vol. 2 Inference and relationship, 4th edn (London: Griffin)

—— 1982 *The Advanced Theory of Statistics* vol. 3 Design and analysis time series, 4th edn (London: Griffin)

Kennerknecht S 1982 *NATO AGARD Conf. Proc. Advanced Casting Technology* p 19-1

Kent H M 1977 Unfulfilled needs of NDI of military aircraft *Proc. 45th Meeting of the AGARD Structures and Materials Panel, Voss, Norway*

Kinchin G H 1979 *Ann. Nucl. Energy* **6** 265–8

Kinsler L E and Frey P 1962 *Fundamentals of Acoustics* 2nd edn (New York: Wiley)

Kirwan B 1982 An evaluation and comparison of three subjective human reliability quantification schemes *MSc Dissertation* University of Birmingham

Klemmer E T 1969 Grouping of digits for manual entry *Human Factors* **11** 394–400

Kletz T A 1983 Why do chemical plants leak? *Seminar Abstract*

—— 1984 *Myths of the Chemical Industry* (Rugby: Inst. Chem. Eng.)

Knoor E 1974 Reliability of the detection of flaws and of the determination of flaw size *AGARD Fracture Mechanics Survey* AGARD-AG-176

Krautkramer J and Krautkramer H 1977 *Ultrasonic Testing of Materials* 2nd edn (Berlin: Springer)

Kupperman D S 1983 Developmental techniques for ultrasonic flaw detection and characterization in stainless steel paper presented at the *OECD/NEA Specialist Meeting on Ultrasonic Defect Detection and Sizing, Ispra, May 3–6, 1983* (Paris: OECD Nuclear Energy Agency)

Kupperman D S and Reimann K J 1978 Acoustic properties of stainless steel weld metal *Proc. Conf. on NDE in the Nuclear Industry* (Ohio: American Society for Metals) p 309

Lancaster J F 1984 *Physics of Welding* **15** 73

Lecomte J C and Launay J P 1979 Reproducibility of ultrasonic amplitudes from reference reflectors *Proc. 4th Int. Conf. NDT Techniques, Grenoble, 11–14 September 1979* pp 13–18

Lenoe E M 1979 Structural ceramics: accomplishments – research needs *Army Materials Conf.* **6** 3

Lewis D M 1951 *Magnetic and Electrical Methods of NDT* (London: Allen and Unwin)

Lewis W H, Sproat W H and Boisvert B W 1979 A review of non-destructive testing reliability of aircraft structures *Nondestructive evaluation, Am. Soc. NDT 12th symp., Texas* (Texas: NDT Information Center, Southwest Research Institute)

Lidiard A B 1979 *1979 Berlin Post–SMIRT Conference Seminars Covering the Efficiency of Detection Function and the PISC I exercise* (unpublished notes)

Lidiard A B and Harrop L P 1978 *UKAEA Report* TP 780

Lidiard A B and Williams M 1977 Analysis of pressure vessel reliability *UKAEA Report* TP 658

Lock D L, Cowburn K J and Watkins B 1983 The results obtained in the UKAEA defect detection trials on test pieces 3 and 4 in *Defect Detection and Sizing. Proc OECD/NEA Specialist Meeting on Ultrasonic Defect Detection and Sizing, Ispra, 3–6 May 1983* (Paris: OECD Nuclear Energy Agency) pp 651–668 (and *Nucl. Energy* **22** 357–63)

Lucia A and Volta G 1983 Requirements for NDI reliability as a function of the size and position of defects in RPVs *Defect Detection and Sizing. Proc. OECD/NEA Specialist Meeting on Ultrasonic Defect Detection and Sizing, Ispra, 3–6 May 1983* pp 793–822

Lumb R F, Bosselar H and Baborovsky V M 1978 Ultrasonic inspection of seamless drill casing and line pipe *Mater. Eval.* **36** 57–71

Lunn J M and Mushin W W 1982 Mortality Associated Anaesthesia (Nuffield Provincial Hospitals Trust)

McCormick N J 1981 *Reliability and Risk Analysis* (New York: Academic)

Mackworth N H 1950 Researches on the measurement of human performance *MRC Special Report* No 268 (London: HMSO)

McMaster R C (ed) 1982 *Nondestructive Testing Handbook* 2nd edn 3 volumes (Ohio: American Society for Metals)

Magistralli G, Sevino F and Vallerani E 1978 Evaluation of crack detection methods *European Space Agency Report* ESA-CR(p)-1092

Malpani J K 1976 *Thesis* Vanderbilt University

Marriott D L and Beyers C J E 1983 *Int. J. Pres. Ves. Piping* **12** 63–105

Marshall W 1982 *An Assessment of the Integrity of PWR Pressure Vessels: Summary of the Second Report of the Study Group on PWR Pressure Vessel Integrity* (London: UKAEA)

Matthews R H 1981 Time dependent analysis of coherent fault trees *UKAEA SRD Report* NST/81/145

Matthews R H and Winter P W 1981 Human factors in risk assessment *UKAEA SRD Report* NST/81/150

Mehran F, Müller K A, Fitzpatrick W J, Berlinger W and Fung M S 1981 *J. Am. Ceram. Soc.* **64** C129

Metals Handbook 1970 8th edn vol. 5(B) (Ohio: American Society for Metals)

Mielnik E M 1974 *Mater. Eval.* **32** 19a–30a

Mika K and Hossfield F 1974 *Bucharest Probability Theory Conf.* (Bucharest: Editura Academiei Republicii Socialiste Romania, 1977) p 329

Miner M A 1945 *J. Appl. Mech.* **12** A159

Mizoguchi S, Ohashi T and Saeki T 1981 *Ann. Rev. Mat. Sci.* **11** 151

Moroney M J 1951 *Facts from Figures* (Harmondsworth: Penguin)

Murgatroyd R A, Seed H, Willetts A J and Tickle H 1983 Inspection of Defect Detection Trials Plates 1 and 2 by the Materials Physics Department, RNL *Br. J. NDT* **25** 313–19

O'Brien K R A, Hollamby D C, Bland L M, Glanvill D W and Scott I G 1975 Crack detection capability of non-destructive inspection methods in relation to the airworthiness of aircraft *Proc. 8th Int. Symp. on Aeronautical Fatigue, Lausanne* ICAF DOC 801

O'Connor P D T 1981 *Practical Reliability Engineering* (New York: Heyden)

Ogilvy J A 1985a Computerized ultrasonic ray tracing in austenitic steel *NDT Int.* **18** 67–78

—— 1985b A model for elastic wave propagation in anisotropic media with applications to ultrasonic inspection through austenitic steel *Br. J. NDT* **27** 13–21

Ogilvy J A and Temple J A G 1983 Diffraction of elastic waves by cracks: application to time-of-flight inspection *Ultrasonics* **21** 259–69

Open University 1976 *Systems Performance: Human Factors and System Failures.* Units 7 and 8 of Engineering Reliability Techniques, a third level course (Milton Keynes: Open University)

Orowan E 1949 *Rep. Prog. Phys.* **12** 185

Osborn W C, Sheldon R W and Baker R A 1983 Vigilance performance under conditions of reluctant and non-redundant presentation *J. Appl. Psychol.* **47** 130–4

Packman P F 1973 Fracture toughness/NDT requirements for aircraft design *J. NDT* 314–24

Packman P F, Malpani J K and Wells F M 1976 Probability of flaw detection for use in fracture control plans *Proc. Joint Japan–USA Seminar, Syracuse University, New York, 7–11 October 1974* ed H Miyamoto and T Kunio (Leiden: Noordhoff) pp 129–43

Packman P F, Malpani J K, Wells F and Yee B W E 1975 Reliability of defect detection in welded structures *Reliability Engineering in Pressure Vessels and Piping* ed A C Gangadharan (New York: ASME) pp 15–28

Packman P F, Pearson H S, Owens J S and Young G 1969 *J. Mater. JMLSA* **4** 666–700

Palmgren A 1924 *Verein Deutsche Ingenieure Zeitschrifte* **68** 339

Paris P C and Erdogen F 1963 *Trans. ASME, J. Bas. Eng.* **85** 528

Parry G W 1979 Regeneration diagrams *UKAEA SRD Report* SRD-R 143

Parry G W, Shaw P and Worledge D H 1979 Tolerance–confidence relationship and safety analysis *UKAEA SRD Report* R 129

Parry G W, Teague H J and Winter P W 1981 Uncertainty in probabilistic risk analysis *IAEA Report* Cn-39/89

Parry G W and Winter P W 1980 The characterisation and evaluation of uncertainty in probabilistic risk analysis *UKAEA Report* SRD-R-190 (and *Nucl. Saf.* **22** 28–42)

Parry G W and Worledge D H 1978 The use of regeneration diagrams to solve component reliability problems *Nucl. Eng. Des.* **45** 271–6

Peck J C and Miklowitz J 1969 Shadow zone response in the diffraction of a plane compressional wave pulse by a circular cavity *Int. J. Solids and Struct.* **5** 437–54

Pettersson B and Axen B 1980 *Thermography* (Stockholm: Swedish Council for Building Research Publications)

Pierce M, Hall J A and Ronald T M 1970 *Aeronaut. Astronaut.* **62**

PISC Plate Inspection Steering Committee 1979 *CEC Report* EUR 6371 vol. I–V (Note that the acronym now stands for Programme for the Inspection of Steel Components)

Pook L P 1983 *The Role of Crack Growth in Metal Fatigue* (London: The Metals Society)

Poulter L N J, Rogerson A, Willetts A J and Dyke A V 1982 Inspection of Defect Detection Trials Plate 4 by the Materials Physics Department, RNL paper presented at *UKAEA Symp. on the UKAEA Defect Detection Trials, Silver Birch Conference Centre, Birchwood, Warrington, 7–8 October 1982* (Warrington UKAEA)

Power H 1664 *Experimental Philosophy* London

Pullin J (ed) 1981 *The Innovative Engineer. 125 years of The Engineer* (West Wickham, Kent: Morgan-Grampian)

Rasmussen N C 1981 The application of probabilistic risk assessment techniques to energy technologies *Ann. Rev. Energy* **6** 123

Rau C A and Besuner P M 1981 *Phil Trans. R. Soc.* A **299** 111

RCOG (Royal College of Obstetricians and Gynaecologists) 1984 *Report of Working Party on Routine Ultrasound Examination in Pregnancy* (London: RCOG)

Reddy D P 1984 (ed) *US DOE Report* DOE/SF/01011-T25, vol.1; DOE 840048 08

Reed A 1985 The warning system that began with a blip *The Times* 26 February p 19

Reed R C 1968 *Train Wrecks* (New York: Bonanza)

Reisland J and Harries V 1979 *New Sci.* 13 September p 809

Reynolds W N and Wells G M 1984 Video compatible thermography *Br. J. NDT* **26** 40

Richards B P and Footner P K 1983 *GEC J. Res. Dev.* **1** 74

Rickover H J 1963 *Quality, the Never Ending Challenge* (Philadelphia: ASTM)

Rogerson A, Poulter L N J, Dyke A V and Tickle H 1984 The inspection of DDT Plate number 3 by the Materials Testing Group, Risley *Br. J. NDT* **26** 20

Rogerson J H 1983 Defects in welds their prevention and their significance in *Quality Assurance of Welded Construction* ed N T Burgess (New York: Applied Science Publishers) p 115

Lord Rothschild 1978 *Risk* (BBC Richard Dimbleby Lecture)(reprinted in *Atom* **268**, 1979)

Royal Society 1978 *Risk Assessment* (London: Royal Society)

Rozina M V and Yablonik L M 1975 Some reliability indices of nondestructive testing and methods of increasing them *Sov. J. NDT* **11** 729

Ruby L 1984 *Am. J. Phys.* **52** 76 (see also *Am. J. Phys.* **45** 380, 1977)

Rummel W D and Rathke R A 1974 Detection and measurement of fatigue cracks in aluminium alloy sheet by NDE techniques in *Prevention of Structural Failure* ed T D Cooper, P F Packman and B G W Yee. (American Metals Society) p 146

Rutter J E, Wiederhorn S M, Tighe N J and Fuller E R 1979 *Army Mater. Conf. Ser.* **6** 503

Sackman H 1975 *The Delphi Technique* (Lexington Books) (see also *IEEE Project 500 Reliability Handbook*)

Salthouse T A 1984 The skill of typing *Sci. Am.* **250** 94

Saporta G 1978 *OECD Report* SINDOC(78)85

Savage W F 1978 in *Weldments: Physical Metallurgy and Failure Phenomena* ed R J Christeffel, E F Nippes and H D Solomon (New York: General Electric Company) p 1

Savot L 1627 *Discours sur les Médailles Antiques* Paris

Schädel J 1983 *Solid State Devices 1983* (Inst. Phys. Conf. Ser. 69) pp 105–20

Schulze W D 1980 in *Societal Risk Assessment: How Safe is Safe Enough?* ed R C Schwing and W A Albers (New York: Plenum) p 217

Scott P M and Tomkins B 1981 Some thoughts on establishing design and inspection codes for corrosion fatigue in *Proc. IAEA Specialists' Meeting on Subcritical Crack Growth, Freiburg, FRG, 13–15 May 1981* ed W H Cullen (Washington, DC: Nuclear Regulatory Committee, 1983)

Scott P M, Tomkins B and Foreman A J E 1983 Development of engineering codes of practice for corrosion fatigue *J. Pres. Ves. Technol.* **105** 255–61

Scruby C B and Wadley H N G 1982 *Proc. 5th Conf. on NDE in the Nuclear Industry, San Diego* (Ohio: American Society for Metals) p 309

Segal Y, Notea A and Segal E 1977 Chapter 9 of *Research Techniques in NDT* vol. 3 ed R S Sharpe (London: Academic)

Seminov A P, Gorbachev V I and Volkov A V 1973 Investigation of the probability of detecting defects by radiation methods *Sov. J. NDT* **9** 739

Serabian S 1982 *Mater. Eval.* **40** 294–8

van der Shchee and Bijlmer P F 1975 Critical survey of methods in *Nondestructive Practices* vol. 1 ed E Bolis (Neuilly sur Seine, France: AGARD) pp 93–128 (AGARD AG 201)

Shcherbinskii V G 1976 *Sov. J. NDT* **12** 224–6 (translated from *Defektoskopyia* **2** 137–9 (1976))

Shraiber D S 1971 Estimating the reliability of nondestructive methods of testing production quality *Sov. J. NDT* **7** 441

Sih G C (ed) 1977 *Mechanics of Fracture* vol. 3 Plates and shells with cracks. A collection of stress intensity factor solutions for cracks in and shells (Leiden: Noordhoff)

Silk M G 1977 Sizing crack-like defects by ultrasonic means in *Research Techniques in NDT* vol. 2 ed R S Sharpe (London: Academic) ch. 2

—— 1978 Estimates of the magnitude of some of the basic sources of error in ultrasonic defect sizing *UKAEA Report* AERE-R 9023

—— 1980 Ultrasonic techniques for inspecting austenitic welds in *Research Techniques in NDT* vol 4 ed R S Sharpe (London: Academic) ch. 11

—— 1981 The propagation of ultrasound in austenitic weldments *Mater. Eval.* **39** 462

—— 1982a Defect detection and sizing in metals using ultrasound *Int. Met. Rev.* **1**

—— 1982b Ultrasonic diffraction as a tool for inspection *Proc. Eurotest Conf. on New Trends in NDT, Brussels* (Brussels: Eurotest)

—— 1984a *Ultrasonic Transducers for Nondestructive Testing* (Bristol: Adam Hilger)

—— 1984b The use of diffraction based time-of-flight measurements to locate and size defects *Br. J. NDT* **26** 208

Singer C, Holmyard E J, Hall A R and Wiliams T I 1957 *A History of Technology* vol. III (Oxford: Clarendon)

Slovic P, Lichtenstein S and Fischoff B 1979 *Energy Risk Management* ed G T Goodman and W D Read (New York: Academic) p 233

Smedley G P 1981 Material defects and service performance *Phil. Trans. R. Soc.* A **299** 7–17

Stephens R W P 1978 Acoustic emission surveillance in the nuclear industry – a review in *Proc. Conf on NDE in the Nuclear Industry* (Ohio: American Society for Metals) pp 191–205

Stojadinovic N D and Ristic S D 1983 *Phys. Status Solidi* a **75** 11

Stoneham A M 1969 *Rev. Mod. Phys.* **41** 82

—— 1981 *Rep. Prog. Phys.* **44** 1251

Stringfellow M W and Perring J K 1984 Detection and sizing of simulated PWR nozzle inner radius defects by the ultrasonic time-of-flight diffraction technique *Br. J. NDT* **26** 84–91

Swain A D and Guitmann H E 1980 THERP: technique for human error rate prediction *Sandia Report* NUREG/CR-1278

Sze S M (ed) 1983 *VLSI Technology* (New York: McGraw-Hill) p 599

Tariyal B K and Kalish D 1977 *Mater. Sci. Eng.* **27** 69

Taylor T T and Selby G P 1981 Evaluation of ASME section XI reference level sensitivity for initiation of ultrasonic inspection examination *NUREG Report* CR-1957 PNL-3692

Temple J A G 1980 Calculations of the reflection and transmission of ultrasound by cracks in steel filled with liquid sodium *Ultrasonics* **18** 165–9

—— 1981 Calculation of the reflection and transmission of ultrasound by cracks filled with solid sodium *Ultrasonics* **19** 57–62

—— 1982 The reliability of non-destructive detection and sizing in *Periodic Inspection of Pressurized Components* (London: Institute of Mechanical Engineers) pp 257–64

—— 1983a Ultrasound reflection from, and transmission through, stratified media with rough interfaces: its relevance to ultrasonic defect detection in *Quantitative NDE in the Nuclear Industry. Proc. 5th Int. Conf. on Nondestructive Evaluation in the Nuclear Industry, San Diego, 10–13 May 1982* ed R B Clough (American Society for Metals) pp 357–61

—— 1983b Time-of-flight inspection: theory *Nucl. Energy* **22** 335–48

—— 1984a The amplitude of ultrasonic time-of-flight diffraction signals compared with those from a reference reflector *Int. J. Pres. Ves. Piping* **16** 145–59

—— 1984b The effects of stress and crack morphology on time-of-flight diffraction signals *UKAEA Report* AERE TP.1067 (and *Int. J. Pres. Ves. Piping* **19** 185–211, 1985)

—— 1985a *Nucl. Energy* **24** 53–62

—— 1985b Sizing capability of automated ultrasonic time-of-flight diffraction in thick-section steel and aspects of reliable inspection in practice *UKAEA Report* AERE R11548 (London: HMSO)

Thomas A, Bailly M and Bieth M 1983 Ultrasonic dimensioning of underclad cracks paper presented at the *OECD/NEA Specialist Meeting on Ultrasonic Defect Detection and Sizing, Ispra, May 3–6 1983* (Paris: OECD Nuclear Energy Agency)

Thompson R B and Achenbach J D 1985 *EPRI Report* NP-3822

Thompson R M 1983 Fracture in *Physical Metallurgy* ed R W Cahn and P Hasson 3rd edn (Amsterdam: Elsevier) ch. 23

Thompson W A 1979 *Oak Ridge Report* ORNL/CSD-45

Tolstoy I 1973 *Wave Propagation* (New York: McGraw-Hill)

Tomlinson J R, Wagg A R and Whittle M J 1978 Ultrasonic inspection of austenitic welds in *Proc. Conf. on NDE in the Nuclear Industry* (Ohio: American Society for Metals) p 64

Tuler F R and Butcher B M 1968 *Int. J. Fracture Mech.* **4** 431

Unwin S D 1984 Binary event string analysis: a compact numerical representation of the event tree *Risk Anal.* **4** 83–7

Uppuluri V R R 1980 *Oak Ridge Report* ORNL/CSD-73

USNRC 1980 *Handbook of Human Reliability* NUREG/CR-1278

Van der Neut 1981 Glass flaw inspection system *US Patent Specification* 4306808

Vesely W E 1970 Time dependent fault tree evaluation *Nucl. Eng. Des.* **13** 337

Waidelich D C 1976 Pulsed eddy current response of point defects *Proc 8th ICNT* paper 3C6

Watkins, B and Cowburn K J 1983 The results obtained in the UKAEA detection trials of test pieces 3 and 4 presented at the *OECD/NEA Specialist Meeting on Ultrasonic Defect Detection and Sizing, Ispra, May 3–6 1983* (Paris: OECD Nuclear Energy Agency)

Watkins B, Cowburn K J, Ervine R W and Latham F G 1983b Results obtained from the inspection of test plates 1 and 2 of the Defects Detection Trials (DDT Paper No 2) *Br. J. NDT* **25** 186–92

Watkins B, Ervine R W and Cowburn K J 1983a, The UKAEA defect detection trials (DDT Paper No 1) *Br. J. NDT* **25** 179–85

Watkins B, Lock D, Cowburn K J and Ervine R W 1984 The UKAEA defect detection trials on test pieces 3 and 4 *Br. J. NDT* **26** 97–105

Weibull W 1939a *R. Swed. Acad. Eng. Sci. Proc.* **151** 1

—— 1939b *R. Swed. Acad. Eng. Sci. Proc.* **153** 1

—— 1951 *J. Appl. Mech.* **18** 293

Welding Institute 1980/1 Size measurement and characterisation of weld defects by ultrasonic testing *Welding Institute Report* 3527/10/80 1980, and *Seminar Handbook on Phases 2, 3 and 4, Coventry, 17 March 1981* (Abington, Cambridge: The Welding Institute)

Wells A A 1982 in *Fitness for Purpose Validation of Welded Constructions* ed R W Nichols (Abington, Cambridge: The Welding Institute)

Wells M G and Hauser J J 1969 Inclusion effects in electroslag remelted high strength steels paper presented at *2nd Int. Symp. on Electroslag Technology, Mellon Institute, Pittsburgh*

Wells P N T 1977 *Biomedical Ultrasonics* (New York: Academic)

West D 1975 A general description of proton scattering radiography *Atom* **222** 54–9

Whapham A D, Perring S and Rusbridge K L 1984 *UKAEA Report* AERE – R 10854 (London: HMSO)

Whitaker J S and Jessop T J 1981 *Br. J. NDT* **23** 293–303

Whittle M J and Coffey J M 1981 *Br. J NDT* **23** 71–4

Wiederhorn S M 1984 *Ann. Rev. Mater. Sci.* **14** 373

Williams J C 1969 Self-paced tracking behaviour *MSc Dissertation* University of Birmingham
—— 1985 Validation of human reliability assessment techniques *Reliab. Eng.* **11** 149–62
Wohler A 1871 *Engineering* **11** 199 *et seq.*
Wolf A K and Green B F Jnr 1957 Tracking studies: 2. Human performance on one-target displays *MIT Lincoln Laboratory Report* 38-31
Wood J 1981 in *Reliability and Degradation* ed M J Howes and D V Morgan (New York: Wiley) p 191
Woodcock J P 1979 *Ultrasonics* (Bristol: Adam Hilger)
Wooldridge A B 1979 The effects of compressive stress on the ultrasonic response of contracting steel-steel interfaces and of fatigue cracks in *Improving the Reliability of Ultrasonic Inspection. Proc. Symp. Br. Inst. NDT, Northampton., 1979* pp 6–18 (and *CEGB Report* NW/SSD/RR/42/79)
Wooldridge A B, Allen D J and Denby D 1982a *Proc. Conf. on Periodic Inspection of Pressurized Components, London 12–14 October 1982* (London: Institute of Mechanical Engineers) pp 109–16 (and *CEGB Report* NWR/SSD/82/0072/R)
Wooldridge A B, Denby D and Allen D J 1982b Predicting and minimizing the adverse effects of austenitic cladding on the ultrasonic inspection of PWR primary components in *Periodic Inspection of Pressurized Components* (London: Institute of Mechanical Engineers) pp 109–16
Wooldridge A B and Duffy P G 1982 The influence of austenitic cladding structure on the transmission of ultrasonic shear waves and its effect on defect detection *CEGB Report* NWR/SSD/82/0006/R
Wooldridge A B and Steel G 1980 The influence of crack growth conditions and compressive stress on the ultrasonic detection and sizing of fatigue cracks *CEGB Report* NWR/SSD/PR/45/80
Wright P 1971 Writing to be understood: why use sentences? *Appl. Ergonomics* **2** 207–9
Wustenberg H and Mundry E 1974 *Proc. Conf. on Periodic Inspection of Pressurized Components, London 4–6 June 1974* (London: Institute of Mechanical Engineers) pp 108–13
Yee B G W, Chang F H, Couchman J C and Lemon G H 1979 Assessment of NDE reliability data *NASA Report* CR-134991/N76-33525
Zambon L B 1971 *Nondestructive Evaluation in Aerospace, Weapons Systems, and Nuclear Applications, Proc. 8th Symp., San Antonio, Texas 24–26 Apr. 1971* (North Hollywood, CA: Western Periodicals Company and Texas: NDT Information Center, Southwest Research Institute) pp 168–71
Zhurkov, S N 1975 *Int. J. Fracture Mech.* **11** 5

Index